W9-ASL-892

AVID

READER

PRESS

Holding Back the River

THE STRUGGLE AGAINST NATURE
ON AMERICA'S WATERWAYS

Tyler J. Kelley

AVID READER PRESS
New York London Toronto Sydney New Delhi

AVID READER PRESS
An Imprint of Simon & Schuster, Inc.
1230 Avenue of the Americas
New York, NY 10020

Copyright © 2021 by Tyler J. Kelley

All rights reserved, including the right to reproduce this book or
portions thereof in any form whatsoever. For information, address
Avid Reader Press Subsidiary Rights Department,
1230 Avenue of the Americas, New York, NY 10020.

First Avid Reader Press hardcover edition April 2021

AVID READER PRESS and colophon are trademarks of Simon & Schuster, Inc.

For information about special discounts for bulk purchases,
please contact Simon & Schuster Special Sales at 1-866-506-1949
or business@simonandschuster.com.

The Simon & Schuster Speakers Bureau can bring authors to
your live event. For more information or to book an event,
contact the Simon & Schuster Speakers Bureau at 1-866-248-3049
or visit our website at www.simonspeakers.com.

Interior design by Kyle Kabel

Manufactured in the United States of America

1   3   5   7   9   10   8   6   4   2

Library of Congress Cataloging-in-Publication Data has been applied for.

ISBN 978-1-5011-8704-9
ISBN 978-1-5011-8705-6 (ebook)

For Marla J. Kinney

Contents

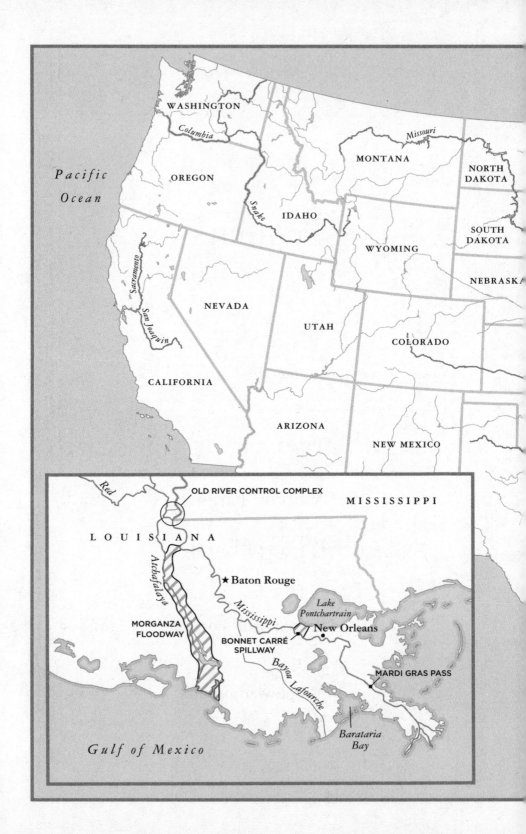

MAJOR INLAND WATERWAYS

0 Miles   500

0 Kilometers   500

MINNESOTA

WISCONSIN

MICHIGAN

MAINE

VT

NH

MA

NEW YORK

CT

RI

INS POINT DAM

IOWA

Upper Mississippi

Allegheny

PENNSYLVANIA

NJ

DAVID LUETH'S
HOUSE

MD

DE

Missouri

Illinois

OHIO

Monongahela

MISSOURI

ILLINOIS

INDIANA

Ohio

WV

VIRGINIA

Kanawha

NSAS

Kentucky

Atlantic
Ocean

KENTUCKY

Arkansas

Inset area below

Cumberland

NORTH
CAROLINA

Ouachita

TENNESSEE

AHOMA

ARKANSAS

Tennessee

SOUTH
CAROLINA

Lower Mississippi

Red

White

Tenn-Tom Waterway

Black Warrior

GEORGIA

Tombigbee

EXAS

LA

MS

ALABAMA

Inset area left

FLORIDA

Gulf Intracoastal
Waterway

Gulf of Mexico

Gulf Intracoastal
Waterway

ILLINOIS

IN

Mississippi

MISSOURI

Ohio

KENTUCKY

LOCK AND DAM NO. 53

LOCK AND DAM NO. 52

OLMSTED LOCKS AND DAM

BIRDS POINT-
NEW MADRID
FLOODWAY

Cumberland

Tennessee

ARKANSAS

Mississippi

TENNESSEE

© 2021 Jeffrey L. Ward

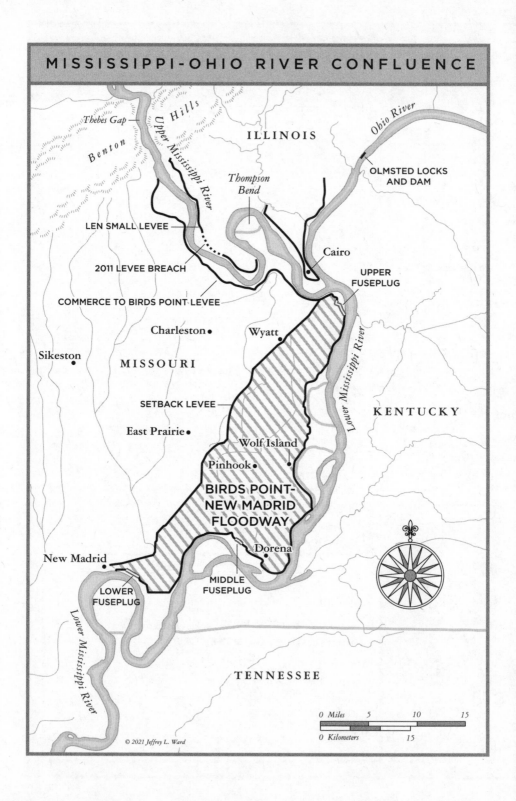

MISSISSIPPI-OHIO RIVER CONFLUENCE

Thebes Gap

Benton Hills

Upper Mississippi River

ILLINOIS

Ohio River

Thompson Bend

OLMSTED LOCKS AND DAM

LEN SMALL LEVEE

Cairo

2011 LEVEE BREACH

UPPER FUSEPLUG

COMMERCE TO BIRDS POINT LEVEE

Charleston

Wyatt

Lower Mississippi River

Sikeston

MISSOURI

KENTUCKY

SETBACK LEVEE

East Prairie

Wolf Island

Pinhook

BIRDS POINT-
NEW MADRID
FLOODWAY

Dorena

New Madrid

MIDDLE
FUSEPLUG

LOWER
FUSEPLUG

Lower Mississippi River

TENNESSEE

0 Miles 5 10 15

0 Kilometers 15

© 2021 Jeffrey L. Ward

Perfecting Nature

David Lueth was riding his lawnmower when the sheriff arrived. The yard he was mowing encircled a white farmhouse where his wife's family had lived for four generations. The green line of the levee was visible just a few fields away. It was the spring of 2011 and the worst flood in history was coming down the Missouri River. In this part of southwest Iowa, all the land between the bluff and the river was under mandatory evacuation, including Lueth's farm. Lueth had moved his crops to safety but was ignoring the evacuation order. After a few obstinate remarks to the sheriff, he agreed to leave.

An unprecedented amount of snow and rain had fallen over the Rocky Mountains and Great Plains. Each of the six massive reservoirs that hold back the Missouri had filled in succession, leaving the U.S. Army Corps of Engineers with no choice but to release the surge of water into the undammed lower river. Up and down the valley, levees blew out or were overtopped. Homes, businesses, and whole towns were inundated. The rushing river rose around Lueth's house and swept into the first floor. The nearby levee was reduced to nubs of earth.

After the water went down, Lueth bought a trailer to live in while he tore out mildewed walls and replaced waterlogged floors. He ordered new farm equipment. He bulldozed the six-foot sand drifts the river had left behind. Rebuilding cost his life savings—$100,000. He had flood and crop insurance, but he couldn't collect until he and other flood victims appealed directly to their senators and the secretary of agriculture.

Lueth has a wide nose and bright eyes quick to smile and quick to cry. He favors round glasses, bucket hats, sandals, and the occasional tie-dye. His shaggy hair has gone gray. Though Lueth, a Christian and an avid bow hunter, is a registered Democrat, he shares a distrust of Big Government with his Republican neighbors. After the flood of 2011, he and his neighbors felt betrayed. The government had promised to protect them from the river, but it hadn't.

Yet by July of 2018, Lueth had canceled his flood insurance. While the Missouri River had been high for several years in a row, he believed that the events of 2011 were an anomaly. Since then, the locals and the Corps had spent a lot of money and time repairing and strengthening the levees. "I'm going to be dead before it all happens again," said Lueth, then sixty-one years old.

Eight months later, it did happen again. Early one Friday morning in March of 2019, the Missouri breached a levee near Bartlett, Iowa, ten miles north of Lueth's house. The water was coming his way, across the Percival Bottoms, a wide plain below the bluffs that enclose the river for almost its entire length. Traffic on Interstate 29, at the base of the Iowa bluff, continued to flow, but the sheriff ordered everyone between the highway and the river to evacuate. Again, Lueth disregarded the order. He had soybeans worth $10,000 stored in a bin beside his house. If the beans got wet, they would swell and rot. He had to move them to high ground.

The flood's cause was obvious enough to Lueth. It had been a cold winter and five days earlier a foot of snow lay on the ground. Then a "bomb cyclone"—a sudden drop in barometric pressure—moved over the Midwest and temperatures rose into the fifties. As all the snow melted, more than two inches of rain fell across several states. The frozen ground could not absorb the water, which ran off into rivers and streams. The Little Sioux, a tributary of the Missouri, rose fourteen feet in twenty-four hours.

Lueth scrambled to buy new flood insurance, but the policy had a thirty-day waiting period. He was too late.

After trucking his soybeans to a nearby grain elevator, Lueth lingered in the house that he had already gutted once. He had recently brought his daughter into his farming business, based in the house he knew he had to leave. The sky was blue and snow geese flew overhead—it was a beautiful day, but Lueth could barely keep from crying. The prospect of abandoning his home and farm made him physically sick. He had thrown up twice that morning. Sandbagging the levee by hand, as he and others had in 2011—groaning and sweating to add a few extra inches of protection—was pointless. The river was rising too fast. The levee, built in the 1940s and rebuilt after 2011, just wasn't high enough. There was nothing Lueth could do.

He left.

The following Monday morning, another levee breached to the south. Parked on an overpass of the now-flooded interstate, he could see his house. It stood on a patch of dry land, but the water was rising. It had inundated a swath three miles wide—from river to bluff—and at least twenty-five miles long. The town of Bartlett was inundated; Hamburg, McPaul, Glenwood, Pacific Junction, Percival—all swamped. Some locals were already saying the damage was too much, that they wouldn't rebuild. In darker moments, Lueth too thought, "Screw it, I'm done."

Over the weekend, his sadness had given way to anger. He was certain the U.S. Army Corps of Engineers, the federal agency that manages the Missouri River, could have done more. Why should he and his neighbors be the ones flooded? The massive dams upstream were supposed to control flooding. Why were they dumping water on him? Why weren't people upstream suffering like this? Were rich South Dakotans, with their summer homes on the lakes, being spared at his expense? Or was it the ice fishermen? He'd heard that the Corps had to release water from the nearest dam or risk bursting it. If the dam wasn't helping, Lueth thought, let it collapse. It couldn't be any worse. Confused, unsure about tomorrow, he prayed and prayed.

Cut off from his own property, Lueth helped his friends and neighbors. While moving farm equipment to higher ground, he saw hundreds of deer running across a flooded field, flushed from the woods by high water. Watching the terrified animals fleeing, as he and his friend also hustled to move, Lueth felt even more depressed. It seemed obvious to him that the river had changed, while the levees and dams built to manage it had not.

"We can't do anything about climate change to make it stop," he said, "but what we can do is to start to build our infrastructure to adapt to it." If the government wanted to use his land for a floodway, they could buy it at market value: $8,000 an acre. If not, they needed to stop the flooding. Something had to be done. "It's time to change. Now."

The next day the water reached Lueth's house. He managed to hop aboard a helicopter with a congressional aide who was surveying the damage. From the air, he snapped a picture of his home: a white box with a peaked roof, a few trees, and, where the grass and driveway and backyard should have been, nothing to the horizon but steel-gray water.

The Missouri River spreading out to fill its floodplain would have looked familiar to Meriwether Lewis and William Clark. The river was doing what it had done periodically for thousands of years, flooding the valley from bluff to bluff. For a mere century—a millisecond in river time—America's engineers had toiled to tame this stretch of the Missouri, to hem it in with levees and dams and dikes. The river was simply asserting a prior claim.

When Lewis and Clark traveled up the Missouri in July of 1804, they camped on a willow-covered island just a few miles from where Lueth's farm would one day lie. The men shot two deer, observed a great number of wild geese, and gathered chokecherries by the handful, pleased to add them to their whiskey barrel. Clark concluded that evening's journal entry with the words "sand bars thick always in view." He knew that the increasing tumult was evidence of the Platte River, pouring into the Missouri a few dozen miles upstream. The Platte carries the sands and silts of the eastern Rockies and has always made the area changeable and flood prone.

In places, the river Lewis and Clark traversed was two thousand feet wide. In others, it was so shallow that boats had to be dragged along its bottom with ropes. Sometimes logjams choked its channel. Banks caved in. Islands emerged and were washed away. The Missouri's personality suited the region's inhabitants: the Otoe, Kansa, Sioux, and Omaha Tribes; the beaver, elk, mosquito, and badger; the grape, hickory, ash, wolfberry, and cottonwood. Not so the land's new claimants, the white government that employed Lewis and Clark. Though the land they explored was beautiful and abundant, it wouldn't do for their purposes.

The leaders and citizens of the recently united states intended to heed God's first words to humankind in the Book of Genesis: "Be fruitful and multiply, and fill the earth and subdue it." Most, if not all, Western thinkers at the time accepted this fundamental worldview without question.

The purposes of Lewis and Clark's expedition, laid out by Thomas Jefferson in a letter to Lewis, his personal secretary, were transparently mercantile and colonial. Jefferson wanted to compete with Britain in the fur trade, and he wanted to acquire land from Native Americans. Jefferson hoped that indigenous people would take up agriculture; once settled on farms, he reasoned in a letter to Congress, they wouldn't need vast woodlands and prairies to hunt in, and these lands could be occupied by whites. On Jefferson's instructions, Clark catalogued town sites, streams suitable for mills, potential farmland, timber, and coal and metal deposits. And he kept meticulous notes for future river navigators.

Advanced civilizations have always been obsessed with controlling water. A river that overflows its banks and changes its course at will is no place for the immobility of bridges, roads, factories, or power plants, no place for the assumptions of concrete and steel. You cannot stake out a modern farm beside it. You cannot navigate its wandering channels with barges or ships, or trust its waters to run through electrical turbines. Capitalism requires predictability, and private property requires boundaries. These concepts are foundational to the way of life that Jefferson personified and codified. In the service of capital and private property, the country's rivers are today held in place by trillions of dollars' worth of concrete, metal, stone, and earth.

The monumental task of taming America's rivers falls to the U.S. Army Corps of Engineers, a mostly civilian federal agency as old as the republic. The Corps' mission expanded beyond purely military construction in the 1820s, when it was tasked with improving navigation on the Ohio and Mississippi Rivers. Following several bad floods in the late 1800s, the agency was given the job of preventing floods. By the mid-twentieth century, most of America's major rivers were operating within a man-made design. Dams created reservoirs and enabled navigation, and levees protected low-lying settlements from high water. Events once written off as acts of God—droughts, floods, crop failures—became the Corps' responsibility. With the rise of environmentalism in the 1970s, the birds, turtles, fish, plants, and mollusks that depend on the rivers were also placed under the Corps' care. Even the rivers' silt was theirs to manage.

Vast tracts of farmland, countless cities large and small, and a sizeable proportion of American industry all rely on what the Corps has built. Particularly essential to commerce is the nation's twelve-thousand-mile network of navigable inland waterways. While the roads and rails may carry more goods by volume, what moves on the waterways is crucial: billions of barrels of American oil and bulk chemicals, nearly all its iron ore, and up to 60 percent of its farm exports—worth about $220 billion annually. If it weren't for cheap river transportation, Brazil and Argentina would have chased American soybeans from the world market long ago.

To protect people and property from high water are twenty-eight thousand miles of levees, more than twice the length of China's Great Wall. Every state has one. On the Lower Mississippi, a single levee system protects four and a half million people living in a swath of the heartland stretching from Illinois to the Gulf of Mexico. Behind this levee are much of the nation's chemical production and oil refining, most of its agricultural exporting infrastructure, and some of its most productive farmland. Since 1928, taxpayers have spent $16 billion on this system, and according to the Corps, it has provided an eighty-to-one return on their investment.

Yet all of this infrastructure has been underfunded for decades. Some of it is deteriorating. Some is sorely inadequate. All of it reflects a set of values, and a set of assumptions about the economy and the climate, that are half a century old and in some ways obsolete.

During the 2016 presidential race, the ungainly word "infrastructure" reached a rhetorical apex—a talking point as popular as taxes, jobs, or "the wall." The one thing the candidates seemed to agree on was that America's roads, bridges, seaports, and railroads were falling apart and needed to be rebuilt. The message resonated with voters on both sides. Hillary Clinton promised to spend $275 billion on infrastructure and to pass an infrastructure bill in her first hundred days. Donald Trump said Clinton's pledge was a "fraction of what we're talking about," and vowed to "at least double her numbers." Clinton said she'd spend $500 billion. Trump raised her to $1 trillion. Neither candidate said much about where the money would come from. Taxes? Tolls? Private equity?

After Trump won, he declared that rebuilding the nation was one of his top priorities. In his Inaugural Address, he said the country had "spent trillions and trillions of dollars overseas while America's infrastructure has fallen into disrepair and decay," and he promised: "We will build new roads and highways and bridges and airports and tunnels and railways all across our wonderful nation."

Two years went by. In April of 2019, Trump met with Senate minority leader Chuck Schumer and House speaker Nancy Pelosi at the White House. The two Democrats came out announcing an agreement to spend $2 trillion on infrastructure. But, in May, Trump walked out of a follow-up meeting with Pelosi and Schumer. The House went on to pursue Trump's first impeachment, and the infrastructure bill died.

A long line of American leaders from both parties has lacked the will, power, or imagination to build what the country needs. In the first half of the twentieth century, American planners got carried away. The government abused its power of eminent domain to seize private property for the alleged public good. Too many neighborhoods were cut in half by unnecessary highways. Too many projects served only to fatten politicians and their allies. Three decades after the New Deal began its crusade to give armies of the unemployed something to do, Americans seemed to lose their enthusiasm for grandiose public works.

Fifty years later, dozens of unquestionably beneficial projects—from high-speed rail lines to sewers, airports, highways, and broadband internet—have yet to be built. Most American infrastructure is at least a generation old. Not since Eisenhower launched the Interstate Highway System in 1956 has the federal government embarked on a comparable undertaking. The quality of American infrastructure was ranked tenth in the world in 2016—down from fifth in 2002—behind countries like France, Germany, Japan, and Spain. European countries spent, on average, 5 percent of their GDP on infrastructure, while the United States spent 2.4 percent. And U.S. infrastructure dollars were increasingly going toward maintenance, not new construction.

In 2017, the American Society of Civil Engineers gave America's inland waterways, which moved about 14 percent of domestic freight and supported more than half a million jobs, a grade of D. U.S. locks and dams are generally designed to last fifty years, but most are considerably older. On the Upper Mississippi and Illinois Rivers—two of the most highly trafficked waterways—the locks were seventy-five and eighty years old, on average. Like cars on a potholed highway, aged locks slow boats down, decreasing efficiency and increasing costs—either farmers are paid less for their crops, or consumers pay more for their food. When Trump spoke about "inland waterways" during an infrastructure speech on the banks of the Ohio, river shippers were thrilled. President Barack Obama had never even mentioned them.

The civil engineers gave America's dams—the ones that hold back water but don't have locks—a D, too. Most dams are owned privately or by states. The report identified 15,500 dams with "high-hazard potential." More than 2,000 of these were also structurally "deficient." High-hazard meant people would die if the dam gave way. Deficient meant that was probable. The cost of rehabilitating all American dams was estimated at more than $64 billion. Fixing just the high-hazard dams would cost almost $22 billion.

America's levees also earned a D. Roughly half the country's levees are privately owned, while the other half are overseen by the Corps. Government levees keep water from flooding $1.3 trillion worth of property, according to the civil engineers, including three hundred colleges and universities, thirty professional sports venues, and one hundred breweries. More than a third of levees were rated "unacceptable" by the

Corps. As with dams, the worst of these were private. The civil engineers estimated that $80 billion was needed over ten years to maintain and improve the country's levees.

Though politicians and the public often overlook the inland waterways, they cannot continue to ignore them. America's roads and railways cannot double or triple their capacity to make room for the goods shipped by river. The farms, factories, cities, and towns behind levees cannot pick up and move. Once taken on, the awesome responsibility of taming America's rivers cannot ever be renounced, unless the nation wants to lose much of what it has built over the last century.

The American people discovered the cost of mismanaging a river in 2019, when the flooding Missouri wreaked more than $3 billion in damage in three states. The Army Corps did not create this disaster. Neither did recalcitrant farmers, or even climate change. The cause was the country's reluctance to give back land taken from the river.

The original plan for damming and diking the Missouri was an uneasy mashup of two visions put forth by two competing agencies. The Bureau of Reclamation's plan was compiled by W. Glenn Sloan, the assistant director of Reclamation's Billings, Montana, office. The Corps' plan was drafted by Colonel Lewis A. Pick. Reclamation wanted to generate electricity and irrigate the arid upper basin, and the Corps wanted navigation and flood control in the more populous lower basin. The final Pick-Sloan plan, submitted to Congress in 1944, called for levees set three thousand feet apart from Sioux City to Kansas City, and five thousand feet apart from Kansas City to St. Louis, where the Missouri empties into the Mississippi. However, these widths were ignored by many farmers, who objected to leaving so wide an arable swath vulnerable to the river's whims. Land creation and conservation were among Pick-Sloan's corollary justifications, so the Corps had little recourse when farmers built their own levees much closer to the river than the engineers thought wise. Though this hodge-podge system was repeatedly overwhelmed by the river, it was rebuilt again and again.

After the flood of 2011 receded, engineers with the Corps' Omaha District put forward a proposal to widen the Missouri River's flood-way by moving some levees back from the river. It was, after all, basic hydrology: the wider a river channel, the more water it carried, and the lower its flood would crest. Picture a wide box with low sides, then a narrow box with high sides—the volume is the same. It was the perfect moment for such fresh thinking, since what was in place had just been destroyed.

The Omaha engineers thought the farmers might be persuaded to return some land to the river, in exchange for better protection from its periodic excesses. They took their plan to southwest Iowa and pitched it to David Lueth and his neighbors: The local levee board would buy the strips of land to be occupied by the new levee, and the Corps would build it. The engineers argued that two local power plants would be better defended, as would people and property as much as thirty-five miles upriver. The levee setbacks, they said, would end an expensive and traumatic cycle of destruction and reconstruction.

The locals refused. The levee board couldn't afford it, and besides, watching tillable ground revert to forest and weeds is a moral affront to a farmer. Where an ecologist sees biodiversity, a farmer sees waste and lost opportunity. The Corps could have acquired the land by invoking the right of eminent domain—the setbacks would reduce flood risk across a wide area and could easily be construed as a "public good"—but as one engineer who worked on the plan said, "We're not going to go there. It's political suicide. We haven't even gone there and it's still kind of political suicide."

The levees were rebuilt almost exactly where they were before the 2011 flood.

At the time, Army Corps of Engineers Colonel D. Peter Helmlinger was stationed in Germany, building housing for soldiers. By 2019 the army had promoted him to brigadier general and brought him back home to command the Corps' Northwestern Division, responsible for all operations between St. Louis and the Pacific Northwest, including the entire Missouri River. The flood of 2019 was his problem. Tall and lanky, Helmlinger had a kindly face that seemed poised to deliver bad news. Every breached levee and every flooded town meant another difficult question to answer, another politician to face.

"After every disaster, there's a lot of momentum to do things differently," he said. "If you don't seize onto that moment, interest wanes.
We missed that opportunity in 2011." Helmlinger was hoping to seize
it in 2019.

He'd heard of congressional interest in better flood control on the
Missouri. He was encouraged by governors who told him, "We can't
keep getting flooded," though he knew that governors "have short
memories." They'd said the same thing in 2011. Like previous commanders, Helmlinger knew there was only one solution: Move the
levees back. The volume of water that taller levees could contain was
"a fraction of what you could gain by putting them apart," he said.
"That's just a simple geometry problem." But to move the levees back,
he needed farmers who were willing to give up land—or leaders with
the political will to take it.

Helmlinger was standing in a passageway of the Motor Vessel *Mississippi*, the trig towboat aboard which the Mississippi River Commission
hosts its public meetings. Helmlinger was being groomed to join the
commission and was along for the ride. Created by an 1879 act of
Congress, the commission was intended to unify disparate public and
private interests under a single authority. Throughout the nineteenth
century, civilian and government engineers had offered competing solutions to the Mississippi's perceived problems, and the legally mandated
makeup of the seven-member commission was intended as a compromise. Three commissioners are from the Army Corps of Engineers, one
of whom is always the commission's president; three are civilians, two
of whom must be civil engineers; and the seventh is an officer of the
National Oceanic and Atmospheric Administration.

Though the commission model has proven successful, no such authority rules over the Missouri or any other U.S. river except the Tennessee.
The loveless shotgun wedding of the Pick and Sloan plans had been a
direct response to the threat of a Missouri River Authority, similar to
the Tennessee Valley Authority favored by President Franklin Roosevelt.

The two rival agencies, Reclamation and the Corps, preferred to share power rather than lose it entirely.

Particularly in its early days, the Mississippi River Commission had some prominent detractors. "Ten thousand River Commissions . . . cannot tame that lawless stream, cannot curb it or confine it, cannot say to it, Go here, or Go there, and make it obey; cannot save a shore which it has sentenced; cannot bar its path with an obstruction which it will not tear down, dance over, and laugh at," wrote Mark Twain in *Life on the Mississippi*.

can't fully control nature

Twain's book begins in the 1860s, before the word "infrastructure" was even in the dictionary, when the river ran where it pleased. Returning to the river in the 1880s, Twain saw that the new commission was attempting a few "improvements"—removing dead trees from the channel, lighting buoys to mark it, building jetties at the river's mouth to deepen its channel into the Gulf of Mexico. Twain scoffed. Any effort to "tame" the river was pure hubris, he thought. Who could control—and with what?—a sheet of water a mile wide, pounding toward the sea at one million cubic feet per second? The improvements Twain ridiculed did little to dissuade the Mississippi from flooding, or running dry, or from snaking across its ancient floodplain as it had done since dinosaurs stalked its valley. The Mississippi, Missouri, and the rest of America's big rivers continued to wander. Many of Lewis and Clark's riverbank campsites are submerged under water now, or buried far inland. Steamboats have been unearthed by Missouri Valley farmers plowing their fields. The boats foundered and sank; the river took off in other directions and buried the boats in future farmland.

The Mississippi River Commission eventually became part of the U.S. Army Corps of Engineers. The Corps considers its accomplishments proof that Twain was wrong. The headline of an article on its website reads: "What Mark Twain said about the Corps of Engineers." Its list of Twain quotations—including his assertion that no one can "tame that lawless stream"—is introduced with the Corps' own assertion: that the changes it has brought to the river "were astounding to those who watched it happen."

didn't fully "tame" the river, as floods still happen periodically. we are not in control of nature but rather minimizing certain processes.

These days, the Corps won't say that it "controls" the Mississippi; "tame" it may venture to claim. Twice a year, in spring and fall, the commission's members sail downriver on the Motor Vessel *Mississippi*. At boat ramps, floodwalls, and small-town docks, the general public is invited to come aboard the floating conference center to present what the commission calls "testimony."

The commission's president always begins proceedings with a speech, and at the second meeting of its 399th session, in April of 2018 at New Madrid, Missouri, that duty fell to Major General Richard Kaiser, a plainspoken natural leader who resented comparisons to Buzz Lightyear. A map behind him depicted the river's meanderings over six thousand years, every channel in a different color. It looked like a handful of rainbow spaghetti thrown at a wall.

"If we didn't do what we did to harness the power of this river," Kaiser told the crowd, "you wouldn't be able to live, work, or farm reliably within eighty miles of this thing, because you don't know where it's going next."

In front of a different audience two sessions later—the 401st—Kaiser wasn't quite so self-assured. The flood of 2019 had begun in February and showed no signs of abating. The second half of 2018, he said, was the wettest period in one hundred twenty-four years across the Mississippi Basin. The earth was more saturated than it had been for twenty-five years. At Baton Rouge, the river would be above flood stage for more than two hundred days. The general could have been laying out a case for climate change, only he never put the words "climate" and "change" together.

Kaiser wasn't being evasive. He was a military man, not a politician. He hewed to the data, the facts on the ground. After another one of his meetings, Kaiser found a bench behind the pilothouse and sat down in the hot sun that most passengers were avoiding. He'd taken off his dress-blue jacket, with its pins and service ribbons, and was looking relaxed in a crisp white shirt.

"People like to play God," he said. "Of course we think it has to be man-made, because otherwise we're not powerful. I can't tell you who is causing it, I can just tell you it's changing." Kaiser believed in change, it was there in the data. "Quite often there's a false causation. They're like, 'We never had water here when I was a kid, we got it now. It's because of what you're doing, and it's not because there's more rain.' When, the fact is, there's more rain."

Kaiser was referring to a report by a senior National Weather Service meteorologist working for the Corps' Mississippi Valley Division. To Kaiser, the report exonerated the Corps. His agency wasn't causing the floods, as members of the public and several lawsuits alleged. Rain caused floods, said the report, and it was raining more. The report didn't offer a reason; it only stated that seasonal precipitation averages were up 6 percent in many places, 11 percent in some. "An increase in annual precipitation and a marked increase in spring precipitation have occurred during the past three to five decades," it concluded. "Therefore, the frequency and the magnitude of floods have increased due to more rain and a significant increase in the frequency of very heavy rainfall events."

More rain. More floods.

A few weeks later, in May 2019, Kaiser gave the order to open the Bonnet Carré Spillway, diverting a portion of the Mississippi into Lake Pontchartrain to control the flood at New Orleans. Bonnet Carré had been opened in three of the previous four years. Now, for the first time in its eighty-eight-year history, it was being opened twice in a single year. Kaiser was nearly forced to open the Morganza Floodway, too, which would have flooded thousands of Louisiana acres for the third time in history. The Corps had been holding public meetings with local people—some of whom might be flooded, many of whom were scared they would be—when the river dropped below the Morganza's trigger point thanks to a series of levee breaches along the Arkansas River.

Kaiser's office had recently reassessed its system, he said, and "There are no major changes that need to be made." Though there was more water between the levees, they were deemed tall and wide enough to handle it. If he'd had hard data to the contrary, Kaiser would have acted on it. Yet the phenomenon called climate change seldom supplies such certainty. So it was raining more. The cause didn't matter.

why does something (distruction, death, displacement) have to happen to trigger change? → people motivated only by money when they have t

The politicians and local interests who came to "testify" aboard the Motor Vessel *Mississippi* didn't want to hear about climate change; they wanted Kaiser to protect them from the river. It remained to be seen whether the Corps could protect the public from a threat neither was willing to name.

This is a hypocritical statement. If climate change is one of the reasons for this flooding, how can they not care about it?

PART I

The Lock

Captain David Stansbury of the Motor Vessel *William Hank* had plenty of time to talk and lots to talk about on a gray Saturday morning in September 2016. Rain was in the air and his subject was delay. His towboat was not moving and would not move again anytime soon. Tied to a fleet of barges near Metropolis, Illinois, the *William Hank* was waiting its turn to pass through Lock and Dam No. 52 at the tail end of the Ohio River. No. 52 was the busiest lock and dam in the United States. It was also the most decrepit.

"This is a choke point right here," Stansbury said. Gracious in his demeanor, forceful in his opinions and predictions, the fifty-nine-year-old paced his pilothouse floor, gesturing upriver toward the unseen structure. "The lock," he said, "is ancient and it's basically falling apart."

When barges are backed up here—sometimes hundreds at a time—a company can apply to jump the queue if its cargo is urgently needed. Once, Tennessee Valley Towing, the *William Hank*'s owner, requested "priority locking" for fifteen grain barges from Iowa. Grain barges are commonplace, but without that particular shipment the chickens at a Pilgrim's Pride poultry plant in Guntersville, Alabama, were going to starve. That barge took a nine-hundred-mile, ten-day trip, transiting fourteen locks on three river systems so that thighs, breasts, and drumsticks could stock America's meat aisles. Traditional high-priority cargoes have been trapped here, too: de-icing fluid for airports, coal for power plants, jet fuel, gasoline.

Unlike a levee, which bolsters the bank, keeping dry land on one side and water on the other, a dam bisects a river in direct confrontation with the current. Each dam on the Ohio backs up enough water to create a navigable channel. In a river dammed for navigation, the dams form a series of pools, like steps. Boats climb or descend from pool to pool by moving through locks, water elevators built into the dams. Moving this way, a vessel can climb from sea level to Oklahoma, Minnesota, or the foothills of the Appalachians. From the *William Hank*'s berth that morning, Stansbury could navigate to the end of the Ohio River at Pittsburgh, 943 miles away, and keep on going, up the Allegheny or Monongahela River.

Days before, problems at No. 52 had stopped river traffic completely. Barges backed up for fifty miles in both directions. Captains kept their deckhands busy—cleaning, scraping rust, painting—but the crew still had time to call their girlfriends, play video games, and strum guitars. A delay of thirty-six hours, with boats idling and burning fuel, cost James Marine, the parent company of Tennessee Valley Towing, around $80,000—and dozens of other companies suffered comparable losses.

Spluttering slightly, unsmiling beneath his white mustache, Stansbury struggled to find a dryland analogy. Pick any big city, he began. Now find that city's busiest freeway, say New York's Interstate 95. "What would happen if both lanes of I-95 were completely shut down for three or four days?" he asked. "You're talking total gridlock in a major metropolitan area. This is the equivalent."

An eight-hour wait is considered fast behind this lock. Eighteen to twenty-four is standard. Mechanical failures of the lock itself have caused week-long delays. Stansbury once spent twenty-eight days on either side of a broken lock, fourteen on one side, fourteen on the other.

Shipping costs rise and fall in proportion to reliability and speed. You don't ship a bushel of soybeans via FedEx. Many commodities, the building blocks of the American economy, move by barge because it is the cheapest way to transport a lot of anything. Per ton, per mile, barging is also the safest and least polluting mode of transportation. Despite its name, a towboat sits behind and pushes a grid of barges: steel boxes, open or closed, tied together with wires until they form a rigid plank. On the Lower Mississippi, towboats can easily push thirty. On locking rivers, a fifteen-barge tow is standard because it fits into a twelve-hundred-foot lock chamber. It would take 216 railcars or 1,050 trucks to haul that fifteen-barge load.

Stansbury retrieved a clipboard with a printout of the *William Hank*'s manifest. Today he was pushing two tankers of soybean oil, one barge of dry cement, one load of aluminum ingots, three loads of scrap steel, four of iron ingots, one barge of wheat, and another of unspecified grain: thirteen barges carrying 19,200 tons of cargo worth around $6.5 million.

The *William Hank* had picked up this tow near Cairo, Illinois, the night before. At Cairo (which rhymes with pharaoh) the Ohio meets the Mississippi, and vessel traffic from the eastern half of the United States meets and passes traffic coming from the north, south, and west. Not far above Cairo, the water and barges of the Tennessee and Cumberland Rivers flow into the Ohio. Apart from the Lower Mississippi, these final miles of the Ohio are the most critical stretch of water in the American interior. More than 80 million tons of goods—worth more than $22 billion—moved through here in 2016. Presiding over it all is Lock and Dam No. 52, the choke point of a nation.

Soybean importers in Japan may have been gnashing their teeth, but the crew of the *William Hank* was concerned with the logistics of lunch. Saturday was steak day. A deckhand radioed the pilothouse, asking Stansbury where to put the grill. Normally, the crew grilled on the bow, amid the ratchets and loops of wire. Today a tank barge of flammable soybean oil was tied up facing the towboat, so the captain ordered the grill moved to the starboard side. Stansbury would eat his supper when he went off watch at 11 a.m.

Crews work twenty-eight-day hitches, with the next twenty-eight off. Some deckhands compare the job, only half-joking, to prison. Life on a towboat can be grindingly routine. Among the traditions passed down from crew to crew and captain to captain are the meals of the day. Friday fish, Saturday steak, and chicken on Sunday. On the *William Hank*, there was one pilot, one captain, one engineer, three deckhands, and two mates. Each knew exactly how many days remained before he could step onto dry land. Among eating, sleeping, and working, eating was the favorite, which made the cook's job the most important. The *William Hank* was lucky to have a cook from Mobile, Alabama, whose soul food was as good as any served in the upscale restaurants of nearby Paducah, Kentucky. In addition to a fridge full of sugary drinks and a closet full of chips—standard on all towboats—there was always a plate of brownies or donuts in the galley.

It might be a sign of prestige for towboat pilots to be overweight. They have earned the privilege to sit up in the pilothouse, handling knobs and levers, not scrambling around on dirty steel barges in the sun, throwing ropes and whacking wires. Stansbury had been a professional mariner for forty years and had an experienced paunch. Starting out in the Coast Guard, he worked the Atlantic Ocean and the Great Lakes for months at a time. After four years of that, he moved inland to work the rivers and to be closer to his family and home in Knoxville.

As the *William Hank* moved up in the queue, Stansbury decided to find a waiting place farther upstream. A deckhand untied the rope holding them to a stranger's barges parked on the bank. The pilothouse shuddered as the shafts began to spin. Carefully heeling out into the current, Stansbury got back onto the subject of the lock: "If 52 does fail, or one of the other locks fails, and you cut off half the United States from their barge traffic, then you'll see a public outcry. 'What do you mean I've got to pay ten dollars for a box of cornflakes? Are you out of your mind?'

"If that lock collapses," he went on, "I'm going to have to find another job, people are going to suffer, prices are going to go up, prices of bread, fuel—we are very vulnerable. To be the most powerful country in the world and to sit here and watch it fall apart . . ." He trailed off.

[margin handwritten note: Envi/river change can cause a massive chain reaction]

Every minute of every day, rivers like the Mississippi and Ohio are trying to escape their earth and concrete fetters and be free. Though catastrophes are rare, the people who live their lives at this interface—between river and levee, or river and dam—are under permanent pressure.

No one felt this weight more than Luther Helland, master of Lock and Dam No. 52. A compact, stoic thirty-seven-year-old, Helland sat on the front steps of the little white frame house that served as his office. The house faced the river, in a line with the lock's pump house and several other buildings. A rock-covered slope led down to the lock walls, first the twelve-hundred-foot chamber and then the six-hundred-foot chamber. Both chambers had miter gates: two steel leaves that closed to form a V, pushing upstream against the current. Beyond the locks was the dam.

All dams, whether designed to control flooding or facilitate navigation, need a way to hold back water and a way to release water. Most modern dams use Tainter gates, curved sheets of metal mounted on long radial arms. Tainter gates can be raised and lowered with the push of a button. Helland had no such luck. His was a wicket dam, made up of hundreds of wooden panels standing side-to-side across the river. If Helland wanted to hold back additional water, or let it go, he needed to send a boat out into the river so that someone could manipulate a series of hinged wooden structures called bear traps, over near the Kentucky shore, or lower wickets.

Nothing in sight was as tall as the river was broad. The lowness of everything beside that wide river made the sky feel close overhead and made the lockmaster look small. The Illinois shore was bluffy, marginally developed, dotted with towns that were losing population; the Kentucky shore was wooded, subject to flooding, good only for hunting. Drivers on the I-24 bridge, just to the west, could easily miss the most important lock in America.

It had been called the umbilical cord of the U.S. Army Corps of Engineers' Louisville District—a remote outpost, chronically underfunded, physically and politically distant from the district's headquarters. No. 52 had changed little since it was built in 1929. It had not been rehabilitated, or much maintained, for decades. It would be a curious artifact if it wasn't so important.

As lockmaster, Helland was responsible for the facility and for every person, vessel, and load of cargo in his lock's vicinity; responsible for the billions of dollars in goods that passed by each year, as well as for the operation of countless factories, power plants, farms, chemical plants, and refineries that relied on the Ohio River as a water source or a trade route. By extension, he was responsible for the livelihoods of millions of Americans. But with the lock and dam in such bad shape, the task of keeping everyone safe and happy was becoming untenable.

"The lock is kept going with all the bubble gum and duct tape we've got left," said Helland. But, he added, "We're running out. She's deteriorating so fast it makes it hard to keep up."

The concrete walls sat on wood pilings driven into the sand of the river bottom. Some pilings were so rotten that only the pressure of the river water kept the walls standing. The lock gates operated hydraulically, but the pipes that delivered the fluid were paper-thin, leaked frequently, and were almost too fragile to patch. Metal throughout the structure flaked and rusted; concrete crumbled and cleaved. Holes and cracks that would have caused any highway bridge to be shut down immediately barely rated attention here. Protective railings had broken off long ago and never been replaced. Many of the lock and dam's parts had been manufactured at the on-site blacksmith shop in the days of anvils and mules. Replacements often couldn't be bought at the store. When there was money, parts were custom-made. Otherwise, Helland improvised.

"This is all farmer work," said the lockmaster, who grew up milking cows in rural Wisconsin and Minnesota. On a farm, he explained, you fix everything with a pair of pliers and some wire. Helland liked to work with his hands, liked the sound of machinery, the hiss of steam, the feel of metal, and the smell of diesel. After high school, he'd spent several years as an army welder and machinist. When he got out of the army, he joined the Corps in Minnesota. After a round of layoffs, he took a job as a prison guard in Western Kansas. He found his way back to the Corps, working at a lock near Chicago, then moved to the other end of Illinois with his wife and five kids for the job at No. 52. Veterans were given preference when the Corps was hiring, and most lockmen had seen combat. The work was considered a relief, maybe even a reward, after the rigors of the armed forces.

Not for Helland. "This is more stressful than when I was in the military," he said. Since becoming lockmaster, he had suffered two "almost heart attacks." His doctor blamed them on stress. Helland worried about his lock every day. In 2015, a family of six drove their pleasure boat over the dam. The boat flipped and four people drowned, including a young girl and her little brother. No. 52 had none of the brutalist towers or overhead cranes that characterize newer dams. Viewed from upstream, the river seemed to stretch on uninterrupted, like an infinity pool. Pleasure boats sailed over the top regularly. Somehow, most escaped injury.

The lockmaster himself had gone over three times. Once, two boats were working on the dam when a cable holding them in place snapped. In full reverse, Helland's 700-horsepower boat couldn't outrun the current and was sucked over the waterfall created when the partial dam

nature is powerful

constricted the powerful river. One lockman in the boat's crew pulled out his cell phone and started dialing. "Why are you on the phone?" a second yelled. The first replied that he was calling his wife to tell her he loved her. Helland, in the driver's seat of the boat's crane, landed upright on a rock. Mercifully, no one was injured. "I really didn't want to go back out there ever again," Helland recalled. In the military, if someone dies on an aerial maneuver, the team does a "confidence jump" afterward. Helland's supervisor sent him out on another boat to see what had gone wrong. "That wasn't a good day," Helland said.

The lockmaster's office was in the first of six houses where employees and their families used to live. At one time, the lock grounds had resembled a small village. Children were born and raised at the lock. No. 52's baseball team played against the team from Lock and Dam No. 53, just downstream. But contemporary lock workers shunned lead paint and preferred air-conditioning; they lived elsewhere. The houses stood deserted, sloughing and sagging. Only the office house had been repainted and maintained. Five hundred and eight miles upriver, at Chilo, Ohio, an almost identical set of buildings at Lock and Dam No. 34 had been converted into the Chilo Lock 34 Visitor Center and Museum. Visitors could study the crude dam, the steam-powered crane, the grim faces of hardworking men in black-and-white photographs—and leave with the impression that such primitive infrastructure was a thing of the past.

Unfortunately for Helland—and the nation—this was not the case.

After finishing his T-bone, Stansbury went to bed. Jackson "Bubba" Walker, the pilot, sat before the stainless-steel levers that controlled the *William Hank*'s rudders, preparing to "make the lock."

A pilot is licensed to operate the vessel and does so 50 percent of the time, while the captain sleeps. But asleep or awake, the captain is always in charge, responsible for every soul on board, as well as the cargo, and any damage or spills or wrecks caused by the vessel or its barges or its crew. (The captain of the *Exxon Valdez* was not on the bridge when it hit an Alaskan reef in 1989, but he was legally responsible for the collision and resulting spill.) If there's trouble, the pilot wakes the captain. Captains can also choose to stop running if, for example, the fog is too

The river acts a vessel to allow this lifestyle and power dynamic

thick or the wind too strong. The owners may want them to push on, but if they hit a dike in the fog, the Coast Guard will investigate *them*, and they could lose their license and livelihood with it. Mark Twain compared these captains to kings, beholden to no one, with a level of freedom and autonomy unknown ashore, and this remains largely true.

Walker, a fifth-generation towboater, could remember floating past No. 52 in the pilothouse of a towboat captained by his father. Now he was forty-seven, with enough years under his belt to more than fill his overstuffed black chair. As the chambers filled and emptied and filled again, Walker smoked cigarettes and squinted through his glasses at the dam. Finally, the *William Hank* was cleared to approach the lock. Walker pushed the throttles forward and the boat chugged slowly toward the narrow, concrete wall.

A thousand feet in front of him, at the head of the tow, a deckhand guided him in over the radio: "All right, Bubba, sitting down here about one thousand below . . . four more feet you be looking at daylight on that long wall . . . points comin' to the good, about a foot or two to the good . . . looking at daylight to the good, you're flat now, you got just a shade over four hundred foot to come abreast . . . about five wide . . ." The voice drawled in rhythmic monotone. "To the bad" meant Walker was going to hit the lock. "To the good" meant he was clear. The pilot had eased in close to the shore early in his approach, to avoid being caught in a slide that might ram him into the bank or the lock. With the power of 4,200 horses, he was driving something longer than the Chrysler Building is tall. He wanted to get it right the first time.

enable people to create speech patterns and slang

Walker maneuvered the *William Hank* snugly into the twelve-hundred-foot chamber, a "temporary" addition built in 1969 to help No. 52 accommodate bigger tows. Instead of a smooth wall, the chamber was made out of poured-concrete cylinders encased in rusty sheet-piles. The setup almost seemed designed to catch the front of a barge. "You can easily get quartered just enough that you can jam up in here and do a bunch of damage," Walker said, laughing. His tow was 105 feet wide, the lock chamber was 110. To park his 1,130-foot craft, he had as much wiggle room as a car in a Walmart parking lot.

Gently tapping the rudder levers and pulling back and forth on the two throttles, Walker steered, came ahead, and stopped in the center of the chamber. An operator in a neon-green vest appeared. He rode a yellow scooter to the end of the chamber wall, got off, and leaned bodily against a long metal lever. The lock groaned. Its sector gears turned, its strut arms stretched out, and its miter gates slowly closed behind the *William Hank*. A lock operator could close the gates and fill or empty the chamber by himself, using the levers and a set of buttons inside two sheet-metal shacks a little bigger than Porta Potties. At a newer lock, it might have taken fifteen minutes to move a boat through, but so much water leaked out of No. 52 that it took an hour.

That made nine, since the *William Hank* had waited eight hours to get here. "This is one of the fastest I've seen it," Walker said. "I have actually sat on both sides of this lock for a week." As water surged into the chamber, slowly raising the boat to the level of the next pool, he studied the dam from the vantage of the pilothouse, four stories above the river. With binoculars, he watched the low line of wickets marching toward Kentucky in a mist of falling water. There was supposed to be a four-inch gap between one panel and the next, but many were missing or out of true. "It's just like holding your fingers up against the water and letting it flow through," he'd said earlier.

The sun had set, the infamous lock was behind them, and Captain Stansbury was back in the pilothouse, guiding the towboat as it churned up the Tennessee River. Walker had taken it up to the Paducah water-front, where the crew exchanged a few barges and took on fuel and water. Stansbury's craft was still 1,130 feet long—247 feet longer than the *Titanic*. To steer, he focused on a blue light on the jack staff, three football fields in front of him on the head of his tow.

A harvest moon rose from behind the woods. Stansbury turned off his spotlights and steered by the moon alone.

At 10 p.m. he was approaching Kentucky Lock and Dam, the last control structure on the Tennessee River and another bottleneck. Over the past 130 years, barges and towboats have gotten a lot bigger, but the chamber at Kentucky Lock was 600 feet by 110 feet, the same size as the first lock chamber ever built on a major American river. As Stansbury neared the lock, ten other towboats were lined up ahead of him.

Because the chamber was small, most would have to break their raft of barges in half and lock through twice. In such cases, a captain would push the tow into the chamber as far as it would go, and—for a standard fifteen-barge tow—a deckhand would separate the first nine barges from the last six, twisting ratchets and tripping winches to pull hundreds of pounds of gear and yards of wire off the couplings. The towboat would then back out, leaving the nine lead barges orphaned in the chamber with the deckhand. After locking through, the powerless raft of barges would be pulled out of the chamber and tied off on the long wall, while the towboat and remaining barges followed—a three-hour process. The *William Hank*, its crew, and the millions of dollars' worth of goods it was pushing, were looking at a twenty-seven-hour wait.

A larger lock chamber has been under construction here since 1998. The projected cost, revised upward several times, tops $1 billion. The U.S. Army Corps of Engineers maintains all the navigation infrastructure on the inland waterways and designs and builds improvements, but money for these projects comes from Congress and from a special tax on gas burned by towboats. Before 2016, there weren't enough gas tax funds to efficiently finance the Kentucky Lock Addition. In 2013 and 2015 it received no money at all. By the latest estimate, the new chamber will be operational in 2023. Until then, tows will have to keep waiting.

"This industry, people just don't see it every day," Stansbury had said earlier. "They don't understand where a lot of these products come from that they use every day. If the public was aware of the commodities we push, we'd probably get more funding. The squeaky wheel gets the grease." It was easy to see why the river didn't squeak. Beyond the pilothouse windows there were no lights for mile after mile. With few towns and fewer homes, no one saw what happened here. Practically the only lights Stansbury passed came from dry docks, coal terminals, and chemical plants. Compared to roads, where the average American drives 11,498 miles a year, rivers are almost invisible.

Stansbury listened with amusement as the lock operator, overwhelmed by unseen voices from unseen towboats waiting in the blackness, called out over his radio to reassure each one of its place in line. After securing his spot, the captain nosed the *William Hank* into the bank for the night. Throttles a touch forward, rudders turned just a little, he could wait indefinitely without tying up. Fog rolled in, and soon he couldn't see the head of the tow. Stansbury had stood the front watch. His six hours were almost up. Walker was awake, brewing coffee under a dim red light. It could take five days to move one hundred miles on these stretches of the Tennessee and Ohio Rivers. Stansbury would sleep six hours and be back on watch before his boat moved again.

*   *   *

On the last day of August 1803, Meriwether Lewis left Pittsburgh with eleven men on his way to join William Clark for their expedition up the Missouri and on to the Pacific. Lewis's early journal entries focused on two subjects: fog and riffles, shallow places where boats ran aground. Riffles were such frequent causes of distress that they were named and dreaded up and down the river. On the third day, the party set off at sunrise and, within two and a half miles, were at their first riffle. "Got out and pulled the boat over it with some dificulty," wrote Lewis. "9 Oclock reched Logtown riffle unloaded and with much difficulty got over   detain 4 hours." He went on:

> Supposed I had gotten over Logtown riffle but find ourselvs stranded again   suppose it best to send out two or three men to engage some oxen or horses to assist us   obtain one horse and an ox, which enabled us very readily to get over   payd the man his charge which was one dollar; the inhabitants who live near these riffles live much by the distresed situation of traveller   are generally lazy   charge extravegantly when they are called on for assistance and have no filantrophy or contience

The party had made ten miles the day before and the next day made only six, so frequently did they have to unload and drag their fifty-five-foot boat through the shallows. The natural Ohio River was replete with waterfalls, rocks, riffles, and shoals. It took Lewis two and a half months to travel *downstream*, from Pittsburgh to the Mississippi River. Though more efficient than hauling people and goods overland, river travel was like a roller-coaster ride with speed bumps. Once set upon her waters, a vessel was at the river's mercy.

Timothy Flint, a Presbyterian minister from Massachusetts, floated down the Ohio with his family in 1815 on the way to the Missouri frontier. To get to the river, Flint first had to make the arduous journey over the Appalachian Mountains by road. There he encountered teamsters, ancestors of modern truckers, driving wagons between Philadelphia and Pittsburgh. "They devote themselves to this mode of subsistence for years, and spend their time continually on the road," wrote Flint, in *Recollections of the Last Ten Years*, a selection of his letters, published in 1826. "They seemed to me to be more rude, profane, and selfish, than either sailors, boatmen, or hunters, to whose modes of living theirs is the most assimilated. We found them addicted to drunkenness, and very little disposed to assist each other."

The United States, then a small, one-coast republic, was expanding into the Ohio River Valley after the Revolutionary War. Just across the divide from the "civilized" world of the thirteen colonies, Ohio was considered "back woods," a land of wolves, where the virulent vegetation made Flint, a prim New Englander, think of decay and sickness.

Arriving finally in Pittsburgh, Flint noted the "funereal" atmosphere of soot and coal smoke, then settled in to wait for the river to rise. It was fall, and the Ohio was too shallow for many vessels to navigate. Steamboats had been invented a few years earlier but were not yet common. Most travelers floated downstream, poling and rowing at the whim of the current. When they got to their destination, they often sold their boats along with their cargo.

Eventually, Flint booked passage on a flatboat, a one-story building on a small, wooden barge. They were cruising along, shortly after leaving Pittsburgh. The captain, Flint imagined, was indulging in "golden dreams of easy, certain, and great profits," when, with a sudden roar, "the river admonished us that we were near a ripple." Dead Man's Ripple, as Flint later learned.

> The boat began to exchange its gentle and imperceptible advance for a furious progress. Soon after, it gave a violent bounce against a rock on one side, which threatened to capsize it. . . . The owner was

pale. The children shrieked. The hard ware came tumbling upon us from the shelves, and Mrs. F. was almost literally buried amidst locks, latches, knives, and pieces of domestic cotton.

Flint told the story of his voyage as if the boat were not controlled by anyone but God, which he likely believed. He described it "jostling on the rocks," "being driven furiously along chutes," or "stuck-fast" on sandbars. In 1815, it was all a person could do to steer a flatboat, let alone control the river it floated on.

Yet the wide river ran through beautiful country. A farmer, Flint imagined, needn't work hard to raise a crop in its valley bottoms. He passed towns of thousands where "eighteen years before, there had been a solid and compact forest of vast sycamores and beeches." He embraced the frontier ethos of wresting order from nature's chaos, while glorying in the "ancient and magnificent forests, which the axe has not yet despoiled."

Four years later, the United States government fitted out an expedition, under the command of Major Stephen H. Long, to travel down the Ohio, up the Mississippi, then up the Missouri. The expedition's purpose was to collect scientific and military data and assess hindrances to navigation. For Flint, a leisurely traveler concerned with souls, the river had been an indomitable natural force. For Long, it was a resource to exploit and improve. Long designed a custom-built steamboat for the trip. The *Western Engineer* drew only nineteen inches of water, had a bulletproof pilothouse, a bow-mounted cannon, howitzers along the sides, and a unique serpent-like design, intended "to surprise and awe the natives," according to a Corps historian. Lewis and Clark had surveyed this territory, but, west of the Mississippi, its only white inhabitants were fur trappers and traders.

Edwin James, botanist and geologist for the expedition, compiled an account of their journey, which was published in 1823. James provided a detailed assessment of extractable natural resources: salt, coal, lead and iron ores, and forests full of timber. "These forests are now disappearing before the industry of man; and the rapid increase of population and wealth, which a few years has produced, speak loudly in favor of the healthfulness of the climate, and of the internal resources of the country," he wrote.

At Wheeling, Virginia (now West Virginia), the party saw the "great national road," the first federally funded wagon road to cross the Appalachian Mountains, connecting the Eastern Seaboard with the Ohio River. Not yet complete when Flint passed that way, it was one of America's earliest public works projects. The roughest section, from Cumberland, Maryland, to Wheeling, had cost an astonishing $1.8 million. If the government had paid to engineer a road, Long might have wondered, why shouldn't it pay to engineer a river?

The defining impediment to navigation on the Ohio was the Great Falls at Louisville, where the river dropped twenty-two feet in less than two miles, cascading down a series of limestone ledges. All boats took on special pilots who knew how to pass safely between the rocks and rapids. Bigger vessels turned around below the falls; those that went up, James wrote, "ascend the rapids at the time of the spring floods, by the aid of a cable made fast to a tree, or some other object above." He went on:

> A steam boat of about two hundred tons, was taken up, and had nearly reached the head of the rapid, when the cable broke, and the boat swinging round, was thrown against the rocks, in the bed of the river, and placed in such a situation as to render hopeless all attempts to get her off before the next annual rise of the water.

The expedition's chronicler followed that harrowing anecdote with a smug rejoinder: "It is expected that the navigation of this dangerous rapid will soon be rendered more convenient, by canaling, which can be accomplished at a very inconsiderable expense."

Influenced by the British tradition of dead-water canals, early American engineers chose not to wrangle an entire stream into a usable form, but to carve a narrow trough overland or beside such natural rivers as the Delaware, Potomac, and Susquehanna. Developing and maintaining a water-borne trade route is expensive, and in the 1700s—as now—it was done only if foreseeable benefits justified the cost. In colonial times, the value was in moving resources like iron and coal from upland mines and smelters to military and industrial centers on the coasts.

Early canals were crowded with locks, but none climbed or crossed the Appalachians. George Washington envisioned a canal crossing the divide between the Atlantic and Mississippi watersheds, connecting Georgetown with the Ohio River at Pittsburgh, but the optimistically named Chesapeake and Ohio Canal Company never made it past Cumberland. The hard-rock geography of the Potomac Valley proved very expensive to canal through, and construction took decades. The Baltimore and Ohio Railroad beat the canal to Cumberland and to the Ohio. A water route over the mountains was never built. The closest the canal companies came was a series of incline railways, where sectional boats were actually taken apart and loaded onto railcars for a steep climb up and over the Allegheny Mountains and down to a canal on the other side.

In 1825 the Erie Canal opened navigation from the Atlantic to the Great Lakes. This connection between two busy watersheds helped New York grow into the economic behemoth it is today and spurred a canal boom. Canals were built as far west as Chicago, where, in 1848, Lake Michigan was joined to the Illinois River. But by the mid-nineteenth century the railroads had put nearly all the Eastern canals out of business. Carrying exponentially more freight, a steam-powered train could cross, in hours, distances that a canal barge—hauled by mule—took days to cover.

The first commercially viable steamboat had been patented in 1807 by Robert Fulton. His company's flagship, tellingly named the *New Orleans*, was built near Pittsburgh. The Louisiana Purchase had opened up the Ohio Valley to the world. With a thousand miles of fertile valley now connected to an ocean port, the Western United States was poised to grow. In an economic sense, Pittsburgh was where the Mississippi started. As the Upper Mississippi merges with the Ohio it becomes the Lower Mississippi. The nominal distinction is a nod to the dominance of the Ohio, which contributes far more water and far more freight to the combined stream than does the Upper Mississippi. Fulton knew, as did Thomas Jefferson, that New Orleans and the Lower Mississippi would make the Ohio Valley great.

Fulton wanted to make himself great, too, and he applied for the exclusive right to operate steamboats on the Western rivers. The Ohio Valley resisted, but Fulton was granted a monopoly by the territorial government of Louisiana, effectively giving his company total control. As a result, keelboating and flatboating remained dominant until a Supreme Court ruling broke the Fulton monopoly in 1824.

All these innovations and advances in river navigation had been the work of private companies, sometimes supported by states. At most, the federal government might do some surveying or buy stock in a canal company to encourage the project. The government regulated trade, but did little to make trade easier.

A private corporation chartered by the Kentucky legislature managed to complete a canal around the Great Falls. The Louisville and Portland became the most profitable canal in the country after the Erie, and—with little government oversight—the canal company raised tolls repeatedly. But, as the size of vessels plying the Ohio grew, the canal was under increasing pressure to grow, too, while the high tolls—enriching shareholders at the expense of Ohio Valley farmers—remained extremely unpopular. As the population and economic output of the valley soared, improvements like a bigger (and free) Great Falls canal began to look like a public good worthy of federal investment.

A week after taking command of the Continental Army, George Washington was lamenting the lack of engineers needed to supervise fortifications and place artillery. With few in the colonies, Washington brought them in from France, where engineer-officers had been esteemed since the reign of Louis XIV. An act of Congress formally organized the United States' own Army Corps of Engineers in 1802. The act also created the nation's first engineering academy at West Point. Jefferson saw the Corps as a source of knowledge on many subjects, far beyond fort building. He considered navigation to be a matter of national security, and asked the West Pointers to advise him on it, though the construction of improvements was still a state responsibility.

The 1824 ruling that ended Fulton's monopoly did so by curtailing state power on the country's rivers. The decision hinged on an interpretation of the Commerce Clause of the United States Constitution. In a concurring opinion, Justice William Johnson wrote that the Constitution had "so clearly established the right of Congress over navigation, and the transportation of both men and their goods, as not only incidental to, but actually of the essence of, the power to regulate commerce." That same year, Congress—newly populated with representatives from the frontier states of Ohio and Kentucky—sent a bill to President James Monroe expanding the U.S. Army Corps of Engineers' authority beyond military fortifications to include civil works that had commercial value. Monroe signed it, giving the Corps the monumental job of making the Ohio and Mississippi Rivers navigable. States' rights politicians moved to defund the program every few administrations, but once begun, the government's role on the rivers inevitably grew.

The army engineers were among the young nation's few experts on hydraulics, so it was logical for Monroe to make the Corps responsible for navigation, a mission that grew to include dozens of rivers. Incrementally over the next century, the Corps was also asked to defend the country from those same rivers in flood. Still within the army, but with a mostly civilian workforce that has now grown to almost thirty-seven thousand, the Corps' focus has broadened over time, but navigation and flood control remain at its heart.

Before it took over the canal around the Great Falls, the federal government asked Stephen H. Long to remove a troublesome sandbar at Henderson Island on the Ohio. From French engineering textbooks at West Point, Long learned about an Italian method of focusing a river's current using dikes, or wing dams. He decided to try one at Henderson Island: a row of wood pilings, two logs wide, stretching from the shore out into the current. The dike diverted the river's flow, funneling water into a new, narrower course. The natural channel was reconfigured, and the sandbar was washed away. Soon it was deep enough for steamboats to pass, year-round. It was the first federal inland navigation project, and a success.

The government continued building dikes and also began a campaign to remove snags: dead trees lodged, roots down, in the river. Navigators were at the mercy of these submerged lances, which could rip the bottom out of a steamboat, wrote Mark Twain, "as neatly as if it were a sliver in your hand." Snags were so common and so reviled that they had their own vernacular: One that moved up and down in the current was a "preacher" or a "sawyer"; one that stood upright was a "planter"; "sleepers" hid beneath the surface. To remove them, the government commissioned a fleet of bizarre-looking snagboats. According to Leland R. Johnson, a Corps historian,

> Service on Uncle Sam's toothpullers was not a picnic. The crews were decimated by accidental drowning and maimed in the complex system of ropes, chains, and pulleys. Disease also found its way aboard ship at times, and snagboats sometimes limped into port with reduced crews—impaired by disease, or by desertions in fear of disease. Service aboard the Western river snagboats was, said the Chief of the Corps, equivalent to Army combat duty.

Within a few decades of Fulton's first trip on the *New Orleans*, steamboats were everywhere. Nearly six thousand were built along the inland rivers between 1820 and 1880; about three-fourths were constructed in the Ohio Valley. The first "motorboats"—though that term was not yet used—steamboats had the unheard-of power to travel upstream. Flatboats had been poled or dragged against the current, often by hand. "I was compelled to know, to my cost, all about pushing a boat up stream with a pole," wrote Timothy Flint. "Justly to appreciate the value of steam-boats on these waters, one must have moved up them, as long, as dangerously, and as laboriously, as I have done." With their pistons and boilers above the waterline, steamboats could carry passengers and large amounts of cargo while drawing only a few feet of water. By flatboat, the trip from New Orleans upstream to St. Louis took months. A steamboat could do it in less than a week.

Floating down the Ohio on a beautiful spring morning, watching houses slip by on the banks, Flint had realized how much variety and breadth of life the river presented to a lonely homesteader. It was the siren song of the wide world; the embodiment of "Go West, young man," a cliché not yet coined by Eastern newspapermen. Life on the Ohio or Mississippi,

Flint wrote, "should always have seductions that prove irresistible to the young people that live near the banks of the river."

Samuel Clemens was such a young person. Born in 1835, he grew up watching the river and its steamboats from his hometown of Hannibal, Missouri. "When I was a boy, there was but one permanent ambition among my comrades," he wrote in *Life on the Mississippi*. "That was, to be a steamboatman." Twice a day, a packet boat arrived at the town wharf. "Before these events, the day was glorious with expectancy; after them, the day was a dead and empty thing." With gaudy adjectives for a gaudy boat, he described a packet:

> She is long and sharp and trim and pretty; she has two tall, fancy-topped chimneys, with a gilded device of some kind swung between them; a fanciful pilot-house, a glass and "gingerbread," perched on top of the "texas" deck behind them; the paddle-boxes are gorgeous with a picture or with gilded rays above the boat's name; the boiler deck, the hurricane deck, and the texas deck are fenced and ornamented with clean white railings; there is a flag gallantly flying from the jack-staff; the furnace doors are open and the fires glaring bravely . . .

The boy did become a steamboat pilot, one of the elect, whose "pride in his occupation surpasses the pride of kings." On the river, Clemens heard the term "mark twain" shouted by the leadsman as he measured the depth of the river, telling the captain that he had two fathoms, or twelve feet of water, a comfortable depth for a steamboat.

As Twain and his kind swaggered about like kings, the railroad barons were not sitting idle. They built tracks to the Mississippi, tracks beside the Mississippi, tracks across the Mississippi, and finally tracks across the mountains and deserts to the Pacific. In the 1830s, there were twenty-three miles of track in the United States. By 1880, there were ninety-three thousand miles. Passenger traffic began to shift from steamboats to much faster trains. Valuable cargoes that needed timely delivery also preferred the rails. The thoroughly unromantic tracks began "intruding everywhere," wrote Twain.

River commerce all but ceased when the Mississippi became the western front in the Civil War. Steamboats were requisitioned and the river was blockaded. After the war, there was freight to move, but the passengers were gone for good. The last straw for Twain's gingerbread steamers came when "some genius from the Atlantic coast introduced the plan of towing a dozen steamer cargoes down to New Orleans at the tail of a vulgar little tugboat."

Towboats, moving tremendous tonnage at low costs, took the bulk traffic that was too inexpensive and voluminous to move by rail. Returning to St. Louis in 1882, Twain described "half a dozen sound-asleep steamboats where I used to see a solid mile of wide-awake ones!" He appended a eulogy:

> Mississippi steamboating was born about 1812; at the end of thirty years, it had grown to mighty proportions; and in less than thirty more, it was dead! A strangely short life for so majestic a creature. Of course it is not absolutely dead, neither is a crippled octogenarian who could once jump twenty-two feet on level ground; but as contrasted with what it was in its prime vigor, Mississippi steamboating may be called dead.

Just as the industry was dying, the federal government got back to work improving the river. The snag program was restarted, and a new system of channel markers and charts begun. Twain thought the improvements came too late. He savored the irony in the voice of a character called Uncle Mumford:

> When there used to be four thousand steamboats and ten thousand acres of coal-barges, and rafts and trading scows, there wasn't a lantern from St. Paul to New Orleans, and the snags were thicker than bristles on a hog's back; and now when there's three dozen steamboats and nary barge or raft, Government has snatched out all the snags, and lit up the shores like Broadway, and a boat's as safe on the river as she'd be in heaven.

If, as some people continue to argue, rearranging the nation's rivers for barge transportation costs the taxpayer and the environment too much, the 1880s would have been the time to stop doing it. Train tracks reached practically every small town, forest, mine, and factory. Almost all the country's goods could have moved by rail, and what remained of waterborne commerce still had natural rivers to run on. But if rivers could offer viable competition, others argued, rail rates would drop. The shippers and farmers who stood to benefit from making rivers more competitive weren't paying for the work, and commerce had come to be regarded as a federal need. No one but railroad stockholders wanted to see their monopoly grow stronger.

Back in the 1840s, a series of primitive state-owned locks and dams were built on minor tributaries of the Ohio River, allowing navigation even in dry years. Before the Civil War, a dam was envisioned across the Ohio below Pittsburgh. The thriving industrial port was handling much of the nation's coal and iron, but the river wasn't cooperating. "During several months in most years, Pittsburgh's steamboats, towboats, and barges remained landlocked, unable to move until it rained and the rivers rose to boatable stages," wrote the Corps' Johnson. Mills would close, shipments would pile up at the docks, and workers would be laid off. The low-water slumps often extended down the Ohio, to Cincinnati, Louisville, and other cities that depended on Pittsburgh's coal. Johnson continued:

> It was during such a seasonal dry spell and accompanying economic recession in 1871 that Pittsburgh's industrialists and businessmen launched their campaign to secure for the city a reliable harbor with a year-round depth for navigation. Pittsburgh should no longer, proclaimed one local newspaper, be dependent "upon the rain from Heaven for conducting our chief employment."

The iron makers prevailed, and Congress authorized construction of a lock and dam to create a slack-water harbor for the city. Army engineers had traveled to see the dams on the Seine and Yonne Rivers, designed by Jacques Henri Maurice Chanoine, chief of the French engineer corps. When rivers were high, Chanoine wanted to allow large coal barges to pass without the restriction of a lock chamber, so he designed a dam with moveable panels, called wickets. The panels could be lowered to allow vessels of any size to pass during high water. When the water fell, the wickets rose, and boats returned to the lock chamber.

Pittsburgh's coal merchants opposed the dam because it would restrict their seasonal flotillas. To placate them, the Corps supersized Chanoine's concept. The Davis Island Lock and Dam, completed in 1885, was more than one thousand feet long, with a 600- by 110-foot lock chamber, the largest in the world. And this monumental project was not a one-off, but part of the first comprehensive plan to tame an entire American river. Forty-nine more dams were planned. They would make all 981 miles of the Ohio navigable, from Pittsburgh to Cairo. The last of these dams, Nos. 52 and 53, were completed in 1929.

In dry months, many American rivers naturally run low enough to walk across. When commerce scraped bottom, only God was to blame. With the lock-and-dam system, the government could promise to maintain a channel at a specified depth year-round. A minimum depth of nine feet was finally chosen for all inland navigation. Modern towboats and barges are built and loaded with this depth in mind. Shoals, riffles, and snags are the Corps' responsibility now; so are floods and droughts. The trip that took Meriwether Lewis more than two months now takes about eight days by towboat, and one day by rail. No longer dependent "upon the rain from Heaven," river commerce depends on the Corps.

Boats never reclaimed the market share they lost to trains, let alone trucks. Steamboats did indeed die, but towboats did not. Trucks moved six and a half times as much domestic freight, by weight, as trains did in 2017. Pipelines moved almost twice as much as trains, and trains moved twice as much as barges or ships. Very little cargo traveled by air, though what did was extremely valuable. The vast majority of American products destined for export ultimately moved by water, much of that across oceans. Many products will use multiple modes of transportation—from warehouse to truck to train to ship, back to train, truck, and warehouse— before they get where they're going. Because some modes are faster, while others are cheaper, a symbiosis has developed. Trains and trucks surely want more of the river's market share, but if goods ever ceased moving on the inland waterways, there would never be enough room on the rails for the extra freight cars, or on the highways for the extra trucks that would be needed to haul the added tonnage. Not to mention increased shipping costs, which would probably destroy domestic industries like farming and steel production.

The navigation project that began on the Ohio spread first to the Lower Mississippi, then moved to the Upper Mississippi, Illinois, Tennessee, Cumberland, Allegheny, Monongahela, Kanawha, Kentucky, Green, Barren, Pearl, Missouri, Red, Arkansas, Atchafalaya, White, Ouachita, Black Warrior, Big Sandy, Sacramento, San Joaquin, Snake, and Columbia Rivers. Lots of smaller streams were navigable for a time, or for a portion of their length. Brand-new waterways were built, too, including the Tenn-Tom, connecting the Tennessee River to the Tombigbee, for an alternative route to the Gulf of Mexico; and the Gulf Intracoastal Waterway, an inshore canal stretching from Texas to Florida. All were either staircased by dams or funneled and held in place by dikes.

Several rivers have had two makeovers. The Upper Mississippi was rearranged with dikes before it was stacked with locks and dams. The first Ohio River dams were small and numerous; since wicket dams have to lie flat during low water, they can't be very tall, and fifty were required to deepen the entire Ohio to nine feet. Understandably, the wicket design was scrapped when a second generation of dams was authorized for the Ohio. The fewer the steps, the faster a barge can move, so the new dams would be tall. Only nineteen high-lift dams would now be needed to maintain the nine-foot channel. Again the project began below Pittsburgh. Before construction on Nos. 52 and 53 had even begun, Davis Island was blown up and replaced with Emsworth Locks and Dams. The replacement projects proceeded downriver until only the two wicket dams were left, and they have been waiting to be blown up since 2001, when Olmsted Locks and Dam was supposed to be finished.

A year had gone by since David Stansbury, captain of the towboat *William Hank*, had said that Lock and Dam No. 52 was "basically falling apart." Now, the catastrophic failure he feared was beginning to unfold. The Ohio was running low, and Luther Helland needed to raise his dam or commerce would bump to a halt. Helland was managing to maintain the navigation channel as legislated by Congress: nine feet deep and three hundred feet wide. But, said Helland, "It hasn't been dry. If she dries out, I don't think I'll be able to do it." There just weren't enough functioning wickets.

No. 52 was supposed to have more than four hundred of them— four-by-twenty-foot oak panels strapped in steel that, taken together, constituted the dam. When raised, the wickets stood in the river like teeth holding back a tongue. In late winter, spring, and mid-summer, when the Ohio was high, the wickets lay on the river bottom and boats floated over them unimpeded. At No. 52 the dam lay down about 40 percent of the year. No. 53, an almost identical dam twenty miles downstream, was down 60 percent of the year. August to November was the traditional low-water season, but rivers are unpredictable. Lock workers had moved wickets on the Fourth of July, on Christmas Eve, and in February, with ice floes piling up against the dam.

Ten of the dam's wickets were missing when Stansbury passed through in 2016; at that time, a one-wicket hole was a big deal, a five-wicket hole a near catastrophe. A single missing wicket could be replaced for about $20,000, plus a day of work for ten people, under ideal river conditions. To fix a multi-wicket hole, the crew had to wait until the Mississippi forced the Ohio to stand still, then dive in the open river. This might happen once a year. It might not. The only other way was to drop the dam, shut down the river—and then dive. Practically speaking, consecutive broken wickets could not be fixed.

The foundation of the dam was a ledge of concrete on the river bottom called the sill. On the sill, each wicket shared a hingelike footing with the wickets beside it. If one broke loose, the two on either side were destabilized by the constant jostling of the river and were liable to break loose, too, domino-style.

As Helland eased his skiff out from behind the lock wall on a cool September morning in 2017, he knew that at least thirty-nine wickets were missing. If the river ran low, and the missing wickets let too much water through, Helland could "lose the pool." If that happened, the river would revert to its natural state. The sandbars Lewis and Clark dragged themselves over would reappear, and towboats would run aground.

When the government—no longer content with the rain from heaven—took over from God, it put people like Helland in charge. He and his colleagues oversee a vastly complicated natural and mechanical system whose operation impacts global commodity prices and the American GDP. Such a task requires constant effort and vigilance. To keep the rivers tame, the Corps of Engineers must keep dancing this dance forever.

Helland danced with his river every day. He knew her quirks (to him, the Ohio was female), her habits, and her whims. He knew what she was capable of and, especially lately, how hard it was to influence, let alone control, her.

Eyes hidden behind mirror-blue sunglasses, mouth set in a straight line, he piloted the skiff into the current. "We failed," he said. "Our dam failed." He didn't mean boats couldn't lock—at least not yet—but that the dam could no longer operate as designed. His crew was solving problems never seen in the facility's eighty-eight-year history. Each and every wicket had always been raised by hand. But, today, all the missing wickets had forced the crew to raise the dam in a desperately inventive way: backward. It was without precedent, free-form, creative. And Helland wasn't sure it would work.

Steering the skiff into the open river, he could see an undulation visible on the surface, where water rushed over the sill. A few dozen wickets had been raised that morning, constricting the river's flow, funneling it into the unwicketed section of the Ohio. Helland could feel the drop, then the smooth leveling out, as he went over the hump and into calmer water—navigating from the upstream side of the dam, where the lock's boats were kept, to the downstream side, where his crew was raising wickets. He had already stopped downbound traffic. The massive web of interconnected economies that depend on the Ohio River was getting anxious, wondering when the dam would be up.

Helland made for a constellation of vessels, small and vulnerable-looking, across the restless river. He tied the skiff to the *Brookport*, the dam's 750-horsepower towboat, named for a rundown hamlet nearby. Helland crossed the *Brookport*'s deck and boarded the maneuver boat, a square, motor-less, and not-very-maneuverable steam-powered crane-barge dating to 1937. A howling jet of flame heated water in the maneuver boat's automobile-sized boiler—the same kind that used to explode on steamboats, leading to numerous fires and deaths in that romanticized era. The only modernization was its fuel: diesel instead of coal. Past the hissing boiler and the spools of cable, Helland stopped at the steel latticework crane that pivoted on a platform near the barge's bow. Steam snaked down the maneuver boat's side and dissolved in the downriver wind.

Jesse Hall, one of Helland's crew, stood on the edge of the maneuver boat, holding a twenty-three-foot steel rod with a hook at its end. The wicket he was fishing for lay under fourteen feet of murky water. Hall slung the rod underhand, aiming at the dead spot, the center of the eddy. He had to hook the wicket's metal "catching bar." If he hit the eddy just right, he could grab the wicket without a struggle.

A talkative, bewhiskered thirty-nine-year-old, Hall had grown up at the lock, in one of No. 52's houses, during the years that his father, Ron Hall, was lockmaster. Ron kept the houses and landscaping as trig as a military base. Jesse often imagined—out loud—what his father would have said about the lock's sad state. His observations didn't seem to endear him to the crew; during breaks, he often sat alone. But he could hook wickets.

Hall felt the river-bottom landscape through the hook rod in his hands. Wood was soft and spongy. Concrete scraped and scratched. Mud and sand were sticky. Metal reverberated and rang. When he hit metal, Hall rotated the rod to see if the hook was set in the catching bar. If it was, he held fast to the rod. Vibrating in the current, it set his whole arm shaking until his partner attached a clasp to the rod's end. The clasp was at the end of a cable, which ran to the crane. Hall gestured to the crane operator, who raised the boom as Hall let go. If the rod held, the wicket would break the surface in a rush of foam.

The operator pressed his pedals and jerked his levers. With two faint clicks, the wicket came up, water sheeting off its pocked wood. As the cable went slack, the hook rod fell off the catching bar and arced wildly through the air. Chanoine's design allowed wickets to be raised from either end, and the crew liked to raise them from upstream, then let the force of the river flip them over and into their proper positions. This way, any debris or sediment that might have been lodged around the resting wicket was flushed out. Hard-hatted and life-jacketed, the crew watched the hook rod fly, listened for the thunk, and waited as the river flipped the wicket. It stood.

The entire exercise was likely a gross violation of federal regulations. The crane—called a "lever rack"—did not meet modern safety standards. Its antiquated design was potentially lethal: The operator's feet had to constantly depress the pedals, ankles rigid, arms poised to yank the iron levers; if he lifted a foot, the steel boom would come crashing down. But there was no money for a new crane and no other way to do the job. (Mention of OSHA, the Occupational Safety and Health Administration, seemed to be taboo around the lock since so much of what was done there was outright dangerous.) "A good lever rack operator is a nervous wreck," Hall had said. "It's safer."

Traditionally, the dam was raised from the upstream side. As the wickets stood upright, the maneuver boat slid out along their tops, prevented from flying into the open river by a steel cable fastened to the lock wall, and by the force of the Ohio pinning it against the dam. The old way assumed the wickets would stand one after another, without issue. When they did, pros like Hall could pull thirty wickets in thirty minutes. That's how well they knew the old machinery, the old methods; it was something rhythmic, beyond technical, a matter of feel, of art. Between 1929 and now, that's how the most important dam in America had been raised. Producers and consumers of basic commodities relied on Jesse Hall's back and forearms, on the nervous feet of the crane operator. Without muscle and steam power, there would be no transportation on the Ohio River.

Under these conditions, however, with all the missing wickets, Helland's team was forced to raise the dam from the downstream side, because the maneuver boat couldn't cross a gap larger than two wickets. A three-wicket hole—or a five—focused the river's current with such force that it could flip the boat and drown her crew. To raise the dam from downstream, the *Brookport* had to pull the maneuver boat back, move sideways, then forward again every six wickets to get the crane into position. This method was a recent invention, necessitated by the unprecedented number of missing wickets. The one other time they had raised the dam this way, it took thirty hours of nonstop work, instead of the normal eighteen to twenty-four.

Hall hooked three wickets, each on the first try. Then it was Marty Jacobs's turn. Jacobs threw, hooked, gestured upward, then tossed his shoulders back and lowered his chin in a posture of confidence. With a hollow metallic roll, the cylinders of the lever rack engaged. Cables caught, the boom went up. The rod flew off and the wicket rose, clicked into place, was let go, and flipped.

Taking a break a few minutes later, Jacobs told his partner, a newbie, about a wicket hooker named Marvin Fairfield, a small man who didn't look particularly strong. "When I first started," Jacobs began, "I'm fuckin' doubled over. Marvin is settin' back there with a cup of coffee and a cigarette. One-handed, he'd take a drink of his coffee, set it on the wicket behind him, get his cigarette, hook it with one hand, they'd pick it up with the crane. Get a drink of coffee, put the coffee down. He'd been here fuckin' thirty years. He knew how to let the water do the work. You ain't going to overpower that—you're gonna lose. When I see that, I said, 'You and me, we're hooking partners. I'm learnin' that shit.'" Jacobs wasn't a large man, either. He'd been hooking for seven years and was one of the best.

As the wickets rose, Helland stood on the deck of the *Brookport*, walked away, came back to squint at a wicket, turned away again. A wind came up and the sky clouded over. Just watching his team work in this unorthodox way upset him. It had been a horrible year for this dam and its lockmaster. Nothing had cooperated.

Raising recalcitrant wickets was challenge enough, but replacing missing wickets with new ones required hard, physical work in another dimension: underwater. River conditions had seldom been right for the task. The previous year, only one new wicket had been installed. This year, the district was paying a contractor to help replace more. So far, they had *almost* replaced one.

That day, Hall had spent the morning diving, with a man above minding his air hose and communication wire. He was protected from the current by a metal dive box, which faced upstream, walling off a section of wickets from the river's constant pounding. Using a jet of water, Hall sprayed sand out of the footing where the new wicket would stand. All rivers have a distinct geomorphology—the way moving water interacts with a stream's bed and banks over time, changing the shape of the river and the surrounding landscape. Big rivers like the Ohio are always up to something. Where Hall was working, linear sand dunes moved back and forth along the river bottom. After several hours, he was finished. All was in readiness. The contractor picked up the wicket with a crane—then stopped.

Government employees like Helland and Hall will work twenty-four-hour days to get the dam up. They make good money on overtime. They might complain but, as ex-military, they won't stop until the mission is complete. Hall had been deployed to Kosovo. Other lockmen were veterans of the wars in the Middle East. After work one day, when conversation rambled around to vacation destinations, one lockman said he didn't like Florida. "I saw enough sand in Iraq," he said.

Contractors are different. To them, this is a job, not a mission. They don't work if they're not paid, or take unnecessary risks, or stay past closing time. So they stopped, with the new wicket dangling at the end of their crane. Their shift was over. The crane turned, lowered the wicket onto the barge, and the contractors went home.

Months later, Hall was still fuming over the incident. So was Helland, who said what almost everyone who works at the old dams feels: "It. Isn't. Just. A. Dam."

The *Brookport* and the maneuver boat passed a two-wicket hole and set up again on the downstream side of the dam. Helland got in the skiff, and the lockmen watched him veer off toward the overgrown woods of the Kentucky shore. The boat, tiny now, with a little orange-vested figure in it, stopped below the trees and the lockmaster got out.

"Figured out why Helland's floating around in the middle of the river yet?" asked Hall.

"Can't figure it out," replied Randy Robertson. Only half-joking, Robertson added, "I think he's having a nervous breakdown." He called Helland's cell, but it went to voicemail. (Helland was in fact talking on the phone.)

For five years, Robertson had Helland's job, but, because of the stress, his doctor told him he needed to do something else if he wanted to live much longer. A friendly, freckle-faced, redheaded fifty-two-year-old,

Robertson was considered an old man around the lock. He remembered when the Motel 6 was a peach orchard, when all the dam's houses were occupied, and when lockmen would get caught sleeping with one another's wives because of footprints left in the snow. He still cracked dirty jokes like the navy sailor he once was. Robertson grew up on the Kentucky side of the river, where oil is pronounced "all" and the town of Olmsted is "Allm-sted."

When the lockmaster position opened at No. 53, Robertson was relieved to transfer to a dam that lay on the river bottom more. The two old dams were almost identical, but because of its position in the river, No. 53 went up only for a few months a year, sometimes just a matter of weeks. Since there was less action, Robertson and his team often helped out Helland and his crew.

"I won't miss this," Robertson said, looking out at the wickets and the barges passing. "It's been a very good career for someone with a high school education," he said, but the job had a way of using people up.

At that moment, Robertson heard the snap of a hook rod disconnecting from the top of a wicket and a subtle difference in the clatter-roll, the pattern of creak, clash, clang—above the constant rush of the river. He knew from the sound that it hadn't been a successful pull, that the wicket would fall back. Robertson watched Jacobs stoop, trying again. "See how he's bending down," he said. "I can't do that." After several injuries, Robertson's back was an almost constant source of discomfort. With his daughter in college, he dreamed of retiring with his wife to Florida, using his captain's license to do fishing charters, setting his own hours, and working only when he felt like it.

Two-thirds of the wickets were up, and the crew and their boats were working from the Illinois shore, this time on the upstream side. It was the second day of free-form dam raising. There were enough intact wickets for the *Brookport*, the maneuver boat, and a third barge holding a small yellow excavator to slide out along their tops, a foot-high wall of gray, river-rounded oak, visible above the water. On the end of the excavator's yellow arm was a hook, specially made to grab wickets. The excavator, or trackhoe, was the latest in wicket-raising technology, but it couldn't reach straight down, which was why yesterday the guys had to hook by hand.

Then the pressure had been on the hook rod experts, Hall and Jacobs. Now it was on Zach Tirey, who possessed a feel for the levers and buttons of the trackhoe that no one else had. Looking younger than his thirty-five years, Tirey had a round face, sandy-red hair, and a pair of studious-looking glasses. On the lock crew, everyone played a role: Hall the armchair quarterback, Helland the sternly silent type, Robertson the jokester. Tirey was the earnest younger brother.

The trackhoe's arm dipped into the water. The sun was low, the river a flat silver. Using the hook at the end of the trackhoe's mechanical arm, Tirey had to find the wicket's catching bar. If he pulled too hard, or at the wrong angle, the heavy equipment at his fingertips could rip the wicket out of its footing on the underwater sill. Tirey found his wicket, lifted it, nudged it over. It flipped and stood. The boats winched out another four feet. Helland, Hall, and several others sat on deck in folding cloth chairs, watching.

Keith Browning sat, too, noting Tirey's progress on a clipboard. Browning was the work leader, the most senior person at the lock after Helland. Thoughtful and universally respected, Browning was the levelheaded one, the quiet older brother. He had left a higher-paying, more prestigious job with the Corps' Louisville Repair Station to finish his career at No. 52, close to his home and family. Though nearing the government's retirement age, Browning had a healthy head of curly dark hair, a narrow face, and a hangdog set to his shoulders. Careful and observant, he was relied on by everyone, including Helland.

As Tirey pulled up a second and a third wicket, Browning looked to his clipboard. He'd tallied wickets present or missing the last time they raised the dam. Their conglomeration of vessels was a few hundred feet into the river now. As each wicket came up, the current grew stronger, swirling and eddying, trying to suck the men and their boats downstream.

Browning gummed a chew of tobacco, spitting periodically. Helland stood, thumbs hooked to his life vest, exhaling sunflower seed shells. On the trackhoe, Tirey reached into the water and came up with nothing. He pivoted and fished for the next wicket. He pulled it up. It stood, but didn't make the requisite click. Something was wrong. Tirey rocked it, lifted, then nudged it with the back of the hook. The wicket flipped halfheartedly. It came up askew, tilting toward Kentucky.

The next one came up sawing from side to side in the swirling white water, as if it weren't connected to the bottom. The wicket hung, passive, going where Tirey led. He let go and it fell beneath the waves. Browning shouted for him to go for the next one, to see how big the hole was. Maybe the wickets were lying on top of one another.

Helland sat, chin in hand. This gap in the dam, three wickets wide and growing, was not on Browning's list. Morale sank as Tirey raised another broken wicket and another. The crew knew that if the next two or three didn't stand, the dam wouldn't go up that day—or maybe ever again.

A healthy wicket should have a snap to it, responding to the pressures of its superior officer, saluting with a "Yes, sir," or a "Yes, ma'am," in a jerky series of predictable moves: up, bump, bump, flip, thud. These wickets were limp, without form or backbone. Dead.

Each new piece of dam focused the river into a narrower opening, like putting your thumb over a garden hose. The more you cover the nozzle, the stronger the spray. The river was moving at five miles an hour now, mind-numbingly slow for a car, but dangerously fast for a mass of water. If this gap, the hole between standing wickets, proved too big to safely cross, it was unclear how Helland's team would raise the rest of the dam. Somehow they would need to get a boat out into the middle of the pounding river, stabilize it, and, from there, reach down for the rest of the wickets. No one could conceive of how to do this without risking people's lives.

Long-range reinvestment in Nos. 52 and 53—expensive projects like shoring up walls or replacing sills—had never been contemplated; the Corps assumed the old dams would be superseded before they failed. Olmsted, No. 52's mega-dam replacement, was supposed to be finished twenty years ago. As the completion date moved further into the future, the Corps scrambled to keep the old dams limping on, pouring concrete here, redoing wiring there, replacing some hydraulic lines. By 2016, the Corps had identified dozens of failure points between the two old dams, any one of which could totally disable them. Meanwhile, the projected cost of Olmsted approached $3 billion. This hunger for dollars coincided with a series of cuts to the army's civil works budget. With each cycle, Olmsted consumed more of the Corps' allowance. Some half-built projects, like Kentucky Lock, went unfunded for years. No. 52 seemed to receive money only in an emergency.

Spitting shells onto the barge's deck, Helland tried diagramming what had happened to his dam. He drew a row of rectangles on a piece of paper. Dissatisfied, he flipped to another sheet and began again. Browning threw the dregs of his coffee into the water. Looking up from his diagram, Helland stuck a finger into his mouth and pulled an imaginary trigger.

Tirey's trackhoe felt around again, hooked a wicket, and lifted—only to have it slip off the hook and slide feebly into the roiling water.

The boats were thirty feet out into the open river now, unsupported, hanging above seven wickets' worth of rushing river. The force of the Ohio—sucking, pulling—made the three boats dip and bend, straining to tear off their wires and lines. Pushing laterally, the *Brookport* tried to hold the two barges against the dam. Even so, the current had pulled the stern of the maneuver boat several feet off the wickets to which it was tied. The back of the trackhoe barge bucked upward beneath the wires that lashed it to the maneuver boat; its bow, tilting forward, was less than a foot from the water's surface—the whole vessel was warping, trying to pivot around the last raised wicket and fly off down the river.

Tirey hooked another. It got away. He spun the trackhoe around and brought its arm to rest over the barge. He took his glasses off, rubbed his neck, and leaned back.

The crew could hear Helland on the phone with his bosses in Louisville: "Stand fast? Really, sir? We don't have anything to stand on." It wasn't safe to move any farther into the unwicketed river. They winched the boat back to the lock wall. The river rushed on.

"The dam's broke," someone said. "The dam broke us," Browning replied. The banter veered between inane jokes and grim pronouncements. Recriminations and blame percolated to the surface. The mood darkened.

Helland was sitting at the base of the steps leading to the *Brookport*'s pilothouse, where several of the crew sat checking their phones and spitting tobacco into paper cups. "Because, I cut northbound traffic at fifteen point oh," he said, distantly. He was trying to convince himself, and anyone listening, that he had prohibited barge traffic from sailing over the wickets in time—when the river was still fifteen feet deep—to prevent their propellers from sucking the wickets off the riverbed and mangling them. No one responded.

The sun was setting behind the I-24 bridge. Helland summoned Browning, and the two sat, legs dangling over the barge edge, talking out of earshot.

Helland finally gave the order to retreat. The men tied up their boats for the night and walked off onto the lock wall. The district bosses called Helland with a plan for raising the dam, but it required the contractors, who wouldn't start work until the morning. Helland was skeptical. "I'll send you pictures," he'd said over the phone, exasperated. No photograph could convey the crew's feelings of failure as those useless wickets had surfaced, one after another.

In a pool of light on the lock wall, the crew gathered around Helland in a ragged circle. The lockmaster favored two- or three-word sentences in Midwestern idiom. He inspired trust and hard work, but didn't tell his crew more than they needed to know. Now he said they were right to be uncomfortable. They had his permission to sit out the next day's job. Several men clowned around in a discouraged euphoria, not really listening. "Does anybody have any objections about tomorrow?" Helland asked. A negative murmur was the only answer.

The river rose the next day, and its current picked up speed. The contractors refused to go out. Helland's superiors came up with a different plan, but he told them it was too dangerous. He knew his bosses were as frustrated as he was, but as he saw it, their proposal was pushing him to risk people's lives to operate a broken dam. Helland was stuck. He hated the thought of letting the river win.

The Ohio kept falling, and, four days later, the dam was holding back so little water that Helland and his crew were able to safely jump the gap and raise the remaining unbroken wickets. Then, a couple inches of rain fell, and the river rose. The dam caught the water, and boats started locking. No. 52 had been shut down for eight days.

*   *   *

Olmsted is the name of a little railroad town on the Illinois bluff. A lock and dam being built nearby took its name from the town. As the lock and dam eclipsed the town, the name Olmsted took on meanings both tiresome and ominous. Like Three Mile Island, it became shorthand for an arduous story with few winners.

The lock and dam that would come to be called Olmsted was authorized by Congress in 1988. For a quarter century, it dammed no water and locked no boats. It was a 160-acre construction site. It took so long to build that the initial contractor was bought and renamed three times; workers began and ended their careers on the site in that time, and new locks at the Panama Canal—Olmsted's international equivalent—were begun and finished.

In Southern Illinois, where the scenery consists mostly of trees and tugboats, Olmsted was spectacular. Not since Union ironclads were hammered together in nearby Mound City had so much money and labor and attention been focused here.

Viewed from the river, Olmsted was a thicket of cranes, miscellaneous machinery, and workboats. From the land, it began as a parking lot fit for a stadium, leading to a batching plant that produced concrete for a fleet of cement mixers. Hollow sections of dam were built beneath a gantry crane the size of a ten-story building. When a section was complete, the crane would carry it to the river on rails and hand it off to a catamaran barge the size of another ten-story building. The barge would float the section onto the river and lower it into place, dozens of feet underwater. Then a second barge would float the cement mixers over. Through thin tubes, the mixers would fill the shell, turning the section into a solid concrete mass. This was called building a dam "in the wet."

Because the Ohio River at this location can rise and fall as much as fifty feet in a season, Olmsted was designed with jumbo wickets and a jumbo wicket-lifter to raise them. When the water was high, tows would continue to float unrestricted over the lowered dam. Unlike Nos. 52 and 53, Olmsted had an air traffic control–style tower that would not be submerged. A series of giant Tainter gates beside the tower would adjust the flow of the river, so the wickets wouldn't have to be handled nearly as often.

Early on, the project was covered enthusiastically in articles and TV shows with titles like "Extreme Machines" and "Mega Cranes." But as the years passed, words like "billions," "overruns," "delay," and "shutdown" came to dominate the headlines.

Everyone on the river had something to say about Olmsted. The workers laboring at the old locks were especially critical. They watched their dams crack and rust—and their jobs get harder and more dangerous—while billions of dollars poured into Olmsted. A message handwritten on a whiteboard once greeted anyone entering the pump house at Lock and Dam No. 53: "Olmsted obstruction site $2,900,000,000.00+." A clipping from the *Paducah Sun*'s business section was taped to a filing cabinet. The headline, about the impact of a government shutdown, read: "Project's 600 Jobs at Stake." Someone had written "Monkey" in front of "Business" and "cartoon section" after.

"We don't call Olmsted a dam, it's a wannabe," said a lockman named Mike Burton.

Randy Robertson, No. 53's lockmaster, remembered cutting brush on an isolated stretch of riverbank as a new hire in the early 1990s. He was preparing the way for the surveyors, the most preliminary step in Olmsted's construction.

Almost thirty years later, Robertson was preparing to retire, and Olmsted was about to officially open. Even in its trial phases, the megadam held back enough water to render No. 53 superfluous since the old dam was so close to it. No. 52, farther upstream, would have to stay in service longer; it wouldn't be demolished until Olmsted was fully operational. On August 29, 2018, the day before Olmsted's ribbon-cutting ceremony, Robertson presided over a much smaller gathering at the base of No. 53's flagpole, beside the decaying red brick lockmaster's house. For the last time, he lowered the American flag. He folded it military fashion and put it in a box. One other longtime lockman stood with him. That was that: no further ceremony to mark eighty-nine years of service. Lock and Dam No. 53 was officially out of business.

Rising out of the water, the lock walls had always looked a bit like ruins. Now contractors were preparing to blow them up. They were already at work, swarming over the lock walls, pumping fluid out of hydraulic lines. When the contractors arrived, Robertson had been surprised to see them dressed in hazmat suits and wearing respirators. They were working with fluids, dust, and paint that he said his crew had handled for years in their regular clothes.

In his office in a sheet metal building behind the flagpole, Robertson dug out a small plastic container with a smaller container inside. He tapped it, and a spray of fine black particles coated his palm. A few floated into the air. "We've been exposed to asbestos, red lead, PCBs— all that," he said.

Robertson and a few of his colleagues had called their union and filed a complaint with the Corps' health office. The government commissioned its own tests, which confirmed the presence of lead, asbestos, and PCBs on both maneuver boats. Robertson said that the health records of everyone who worked at Nos. 52 and 53 now note their exposure to dangerous levels of several known carcinogens. He believed that, just as it had pushed Helland's team to raise the dam in conditions it knew were unsafe, the government knowingly exposed its workers to these substances rather than repair the dams or shut them down. (In an emailed statement, the Corps' Louisville District said they were aware of the toxic materials, but had no evidence that employees had been exposed, noting that the "presence of materials does not equate to exposure" and asserting that the district had taken appropriate steps to ensure worker safety, including requiring the use of protective equipment.)

"Why is it OK for us, as federal employees, to use this old crap, when you've got contractors working under federal guidelines can't get anywhere near it?" Robertson asked. The black dust was soot from the maneuver boat's boiler. Robertson tried to rub it off his palm. "See how it stays on my skin?" he said. "When I'm lying on my deathbed and I've got cancer, I want the government to take care of my family."

Robertson left his office and got into his truck for a farewell look at what was no longer his lock. Aboard No. 53's maneuver boat, he banged open the boiler's hood, revealing its dusty black guts. Like Helland, Robertson grew up on a farm. Coming to work at the lock as a young man, he found the mechanics familiar. When he first hired in, he recalled, his job was to wriggle inside an opening not much bigger than a Frisbee and caulk the boiler from inside. Above that opening, a honeycomb of thin tubes ran the length of the huge cylinder. Soot lined every tube, the same soot that had been worrying Robertson. It coated the tops of the fluorescent lights. It was everywhere.

Robertson paused at the chair once reserved for the maneuver boat's crane operator. He looked out over rows of shoulder-high gears and the cables extending thirty feet behind them. "I know how to take all this apart," he said, "And it doesn't matter."

Yet Robertson wasn't out of a job. He would be in charge of Olmsted's boats, the master of the fleet. The most important vessel under his care was the giant wicket-lifter, based largely on the maneuver boat but with an arm like a trackhoe. The seats in the wicket-lifter's cab would match the space shuttle's, Robertson said. The technology they worked with was taking a giant leap, from 1920 to 2020. As of tomorrow, no Corps employees would be found at No. 53. They had all been transferred to Olmsted, which, though it had been built by contractors, would be owned and operated by the Corps. The crew from No. 52 was already training with Olmsted's wickets. When their dam was decommissioned, they'd be transferred, too.

Robertson had hoped the Smithsonian would take the maneuver boat, but the vessel was too decrepit and contaminated. He thought the city of Olmsted might take a wicket and a hook rod and add them to its historic train display, but the maneuver boat and most of the old equipment would probably be scrapped. He didn't expect guys like him, who came up on steam, to last long at Olmsted. "We'll fade away within three to five years. Can't see Luther there for long." The change wouldn't just be technological; it would be cultural. The guys at No. 53 used to cook Christmas dinner at the lock when they couldn't take the day off. Since the old locks were so low and the work so physical, the crew operating them interacted with every boat that passed, tossing lines, making jokes, complaining. Once in a while, a deckhand on a passing towboat might hand over a plate of food.

Olmsted's lock operations would be accomplished by a worker seated at a bank of computer screens in an air-conditioned tower. Helland said that at No. 52, he knew if the lock was in working order by the feel of a lever, the hiss of hydraulic fluid, or the groan of a gate. At Olmsted, computers would run everything. Helland would be the assistant lockmaster there, but he couldn't imagine running a lock from an air-conditioned room. "I like to be able to hear it, see it," he said. "I might be miserable, cold, hot, but at least I'll know what it's doing."

Standing there on the barge as the sky clouded over, Robertson seemed less resentful than relieved. "A load of bricks is being lifted off my shoulder," he said—a load that would soon fall on the shoulders of Shane Byassee, a quiet forty-five-year-old from across the river in Hickman County, Kentucky. Byassee had hired into the Corps soon after high school. Now he was Olmsted's lockmaster. "Friday morning, he's going to wake up and go, 'Oh, crap,'" said Robertson. Then the Ohio River would be his problem.

The next day, August 30, 2018, Byassee stood with his wife and teenage daughter beside a large tent set up for the Olmsted ribbon-cutting. Olmsted's lockmaster wore a white shirt, a buzz cut, and a slight smile. His voice was low and smooth, but Byassee wasn't a big talker, a virtue in a job that was going to expose him to pressure from every direction to get the dam up and the lock open. Byassee would have to toe a line stretching to Louisville and then on to Washington. But the machinations of how and why Olmsted was built weren't his problem. His job was to operate the thing.

The ribbon-cutting, in fact, was symbolic. The dam wasn't opening that day. Steel in the gates was cracked. The hydraulic cylinders didn't have enough fluid. During routine raising and lowering, a wicket had broken off from the river bottom, snapping its bolts. No one knew exactly why. And the river might not give them time to find out. It was falling, and would soon be too low for No. 52, now missing more than eighty wickets, to hold pool. Ready or not, Olmsted's dam would have to go up soon.

It was hot under the tent. Some one thousand people sat or stood, waiting for the ceremony to begin—Corps employees, retirees, their families, and a gaggle of contract workers in dirty boots, camouflage caps, and neon company shirts: "AECOM Alberici Olmsted 2018." AECOM and Alberici were the principal contractors that built Olmsted. The neon-shirted attendees looked pleased. So did the Corps' project managers, in dress shirts and name badges. For both groups, Olmsted represented a career's worth of well-paying work.

Luther Helland had said he might not come, but he was spotted sitting near the front of the tent in a blue button-up shirt, staring straight ahead. He wasn't wearing his uniform, but Lieutenant General Todd Semonite, chief of the Corps, wore his, as did Major General R. Mark Toy, commander of the Great Lakes and Ohio River Division—both in combat fatigues. Kentucky Senator Mitch McConnell and Illinois Senator Dick Durbin came by helicopter with Secretary of the Army Mark Esper and his assistant secretary for civil works, R. D. James. Interchangeable in their suits, they arrived late and took front-row seats.

Speakers stepped to the lectern in order of importance. The back-patting and jargon ran together. A man from AECOM called Olmsted "a wonder of modern construction." A barge-industry lobbyist reminded everyone that in 1988, when it was authorized, Olmsted was projected to return eight dollars for every one dollar spent to build it. (When Olmsted would be delivering $24 billion, he didn't say.) Durbin called this "a rare moment, when the river will bend to the hands of man." He and McConnell talked about the jobs created, all the houses people would buy, and the kids who'd be going to college thanks to Olmsted.

Finally, a table was brought up to the podium. In the middle of it was a red circle the size of a pizza. The "button." The officers and politicians bent forward, pretended to press, and froze for the camera. A towboat, waiting behind the lock gate, sounded its horn. Video screens showed the gates slowly opening. The boat chugged forward. It was a mock lockage. The chamber wasn't emptied or filled. The dam wasn't holding back the river, not yet.

After the ceremony, the brass answered questions. Semonite, pugnacious and loud, didn't need a microphone. He was clearly having a great time. Everyone hewed to Olmsted's public relations spin: "Innovation" was used five times in their prepared remarks, "revolution" twice. Delays and cost overruns were the fault of Congress and the "incremental funding" it doled out over too many years. The week before, in a conference call with the press, Semonite had spoken about "revolutionizing" the Corps as if it were a Silicon Valley start-up. He extolled Olmsted's experimental in-the-wet construction. Anything that wasn't "innovative," he seemed to be saying, was slow, old, and boring, including the traditional way of building locks and dams.

When Olmsted was first authorized, in 1988, Congress had approved a project that would cost $775 million and take seven years. This was presented as established fact, though the Corps hadn't yet decided how Olmsted was going to be built. For over a century, the standard method had been cofferdams, temporary watertight structures set into rivers and then pumped free of water. Cofferdams narrow a river's width, hampering commerce, but they give builders a dry construction site. All of the original Ohio River dams were built this way. At Olmsted, a cofferdam was used for its two twelve-hundred-foot locks. That contract was put up for bid in the usual way and work began in 1993.

Mike Braden, the Olmsted division chief, said he was a fan of cofferdams, but the towing industry had tired of them. They took too long. A cofferdam had failed at an Ohio River lock in the 1970s, setting the project back by a year. Pressure had grown to find a new way. Braden called it "one of those emotional things."

The most definitive account of what happened next—which Braden considers "fair and unbiased"—is the Government Accountability Office's 2017 report, *Factors Contributing to Cost Increases and Schedule Delays in the Olmsted Locks and Dam Project*. The GAO is the audit, evaluation, and investigative arm of Congress, and its report tells a markedly different story from the one told by the Corps on that hot August day. According to the report, "in-the-wet" construction and the type of contract the Corps used to build the dam were the "primary contributors to cost increases and schedule delays," not inefficient funding, as Semonite and the others had claimed. The Corps had originally planned to construct Olmsted using four cofferdams, said the report, but "the project was the subject of many studies and reviews seeking to improve on the authorized plan by incorporating innovative design and construction methods." These studies, commissioned by the Corps, showed that it would be cheaper and faster to build the dam in-the-wet; the technique had been used to build tunnels and bridges in a marine environment, but not "to construct a project such as Olmsted in a river environment."

The district decided to do it anyway, citing "lower cost, shorter construction schedule, less impact on navigation during construction, and the potential for fewer negative environmental impacts." At the time, said the GAO, this decision "was not required to undergo an agency technical review or an independent external peer review."

Having settled on a construction method, the dam contract was put out for bid in 2002. The Corps offered a firm, fixed-price contract, asking the builder of the project to commit to a dollar amount and a time frame. The contract was short on details. It didn't tell the contractors how to build a dam underwater, just that "in-the-wet" construction was what the Corps wanted.

No bids came back. The project was too big and the methods too experimental, especially on a river where water levels can fluctuate fifty feet from spring to fall. It seemed that no construction companies could get bonding; their underwriters deemed the project too risky to insure.

In hindsight, Mike Braden said, they should have taken this feedback to heart. Unlike the government, the private sector clearly thought in-the-wet construction could lead to cost overruns and delays. But, even at that early stage, Braden recalled, the Corps was in too deep; it couldn't turn back. The lock and its approach walls had been built. They had a design for the dam, and a contract that had taken more than two years to prepare. "Do you want to shitcan three years of that lead-up and start over?" Braden said. The modeling and the environmental review would have had to be redone. The project would have had to be opened, again, for public comment.

So the Corps didn't change its design. It just put out a new contract for bid. This time, it was a cost-reimbursement contract, essentially a blank check. The contractor would put a number on its bid, but wouldn't be held to that number, or be required to complete the project by a certain date. With cost and time practically unlimited, two bids came back. The Corps chose the lowest.

As construction began, Hurricane Katrina hit the Gulf Coast. The Corps' levees around New Orleans failed, thousands of homes were inundated, almost two thousand people died, and a public relations disaster unfolded. The Corps rushed to put out billions of dollars in contracts to repair New Orleans's flood defenses. Seemingly overnight, Braden said, the cost of barges and cranes tripled. The prices of steel, cement, fuel, and insurance all increased dramatically.

Work proceeded—slowly. The Corps gets its civil works allowance from the president's annual budget, which meant that Olmsted could get money in any given year, or not. In the early 2000s it got very little. Also, in-the-wet construction wasn't possible when the river ran high, so work periodically had to stop.

From 2007 to 2010, said Braden, "We ran smack into the learning curve." The contractor had to figure out how to cast the huge concrete shells, float them, sink them, and fill them. They had to invent the giant gantry crane and the catamaran barge. They had to figure out how to organize the barges, the concrete plant, the cement mixers, and all the other equipment, on land and afloat. "We try to be on the cutting edge, not the bleeding edge," said Braden. "This was the bleeding edge." The first concrete shell wasn't placed underwater until 2010. That was the year, Braden said, that they learned "they'd blow their budget." If a federal project's cost increases by a certain percentage, the responsible agency must inform Congress. Congress can then either cancel the project, or reauthorize it at a higher cost.

The Corps' request to reauthorize Olmsted landed in the lap of Senator Mitch McConnell, who must have known that it would generate controversy. He waited until he had maximum leverage to push it through; he included Olmsted in a bill to reopen the government, ending the shutdown of 2013. The Republican's own Tea Party called the reauthorization fiscally irresponsible, but House minority leader Nancy Pelosi didn't seem to mind. She let the funding stay in the bill. When asked later about McConnell's pet dam, Pelosi said, "Whatever it was, it was not enough to say we're not going to open up government because there's something that Senator McConnell put in about a road or something. I don't know what that was."

That something was increasing the budget for Olmsted from $775 million to $2.9 billion.

Realizing that in-the-wet was the problem, the towing industry asked the Corps if it could abandon its experiment and bring back the reliable old cofferdams, said Braden. In 2012, the Corps studied the idea and, according to the GAO, estimated that Olmsted would be operational by 2022 if built in-the-dry, at a cost of $2.8 billion. Built in-the-wet, it would get going in 2020 and cost $2.9 billion. The Corps opted to continue in-the-wet. If it hadn't, Braden said, Olmsted would have taken not two, but five or six, years longer to complete. Congress signed off on the $2.9 billion.

In the meantime, the towing industry raised the possibility of rehabilitating Lock and Dams Nos. 52 and 53. The Corps studied this, too, said Braden, and concluded that "any day you lock a boat through, consider yourselves fortunate. No reasonable amount of money would keep 52 and 53 viable." Fifty-two was like a wire hanger, said Braden, bent back and forth until it was about to break. It had been bent back and forth for almost ninety years. Rebuilding the dam from the riverbed up was the only solution.

For several years, Braden recalled, there was an "emergency, sky is falling plan" for 52: Rock the river. They'd dump a long pile of limestone over the ruined wickets, creating a new dam on top of the old one. Boats could pass though the lock until the river rose. Then a hole could be blown through the rock to open the river back up. For two summers, barges loaded with limestone were parked on the riverbank just a few miles from Helland's office, awaiting his call.

Andy Schimpf, manager of the Corps' second newest and second most expensive lock and dam, compared Olmsted to the Vietnam War. "Everybody now would concede that the way the country handled the Vietnam conflict was wrong," said Schimpf. "We should have either gone in there and finished it, or never got involved. But, at the time, no one on either side was ever going to concede that we're not doing this right, because nobody wanted another major war, but they also didn't want to pull out and say, 'Boy, we screwed this up.' "

There is something fundamentally American about not giving up, not admitting defeat. Even if it harms everyone involved, slogging on ahead is better than quitting. "To me it was a lot of egos," said Schimpf. High-ranking people had put their names on in-the-wet. Switching to in-the-dry, though it would have saved money, "would have been a major admission on the Louisville District that we screwed this in the worst possible way."

Schimpf worked briefly at Olmsted and later became the rivers project manager for the Corps' St. Louis District. A civilian, he oversaw several locks and flood control projects and handled a budget in the tens of millions. One of his responsibilities was the Melvin Price Locks and Dam, named for a Democratic representative whose district included East St. Louis. Mel Price, which began locking boats in 1989, cost nearly $1 billion and took nine years to build. "I thought that was forever," said Schimpf. Back then, nine years was a very long time for a civil works project to be under construction. Now, Mel Price's construction-cost overruns (less than 10 percent) and timely delivery look "unbelievable," compared to Olmsted, said Schimpf. The difference was planning. Olmsted's open-ended contract left crucial parts of the project undesigned. "At Mel Price, everything was engineered, plans and specs," Schimpf said, "down to a gnat's ass."

Two days before the Olmsted ribbon-cutting, Schimpf was at the National Great Rivers Museum, adjacent to Mel Price, sitting in front of a mural that may have depicted what the landscape outside looked like before the Corps got hold of it.

Schimpf—short, gray hair; clipped goatee—spoke two languages: river engineering and baseball, his other obsession.

The new Busch Stadium, home of the St. Louis Cardinals, was built in two years, Schimpf pointed out. Why? "They had planning." And they had money, from the county and state, a private bond issue, and the team's owner. "I am absolutely convinced: You could build Mel Price in two years, you could build Olmsted. There's not an engineering reason that keeps us from doing that," he said. A new Busch Stadium was also a good investment. The Cardinals are a product, and the new stadium improved sales.

The Corps' relationships with products and customers may be less apparent, but in Schimpf's view, businesses that benefit from navigation on the inland waterways ought to pay more. "We have a partner that's not going to come to the table, no matter what they might say publicly—and why would they?" he asked. That partner is the towing industry. In public, the towboat owners and their lobbyists boast about the money they contribute to waterway projects; in private, they work to make sure that river transportation remains heavily subsidized.

If waterway projects were as beneficial as the Corps claims, Schimpf suggested, their beneficiaries—soybean farmers, river shippers, global grain exporters like Archer Daniels Midland—would pay more for them. When a lock closes unexpectedly, Schimpf said, companies are quick to call him, saying things like, "This is costing me two million dollars a day." "The repair will cost five hundred thousand dollars," Schimpf tells them, "so why don't you fix it and save some money?" Silence on the other end. If the industry ever volunteered to pay more to use the nation's waterways, Schimpf figured, they'd be expected to pay at that rate forever. Such a precedent might even cost $2 million a day.

The current model for funding inland navigation projects was born during construction of Schimpf's own Mel Price. Until then, the federal government had paid the entire cost of every lock and dam. But when it set out in the late 1960s to build Mel Price, a replacement for the aging Lock and Dam No. 26, near Alton, Illinois, the environmental movement was beginning to gain national attention. Environmentalists opposed the project as a threat to the river's ecology. Railroads—in competition with waterways since their inception—decried the project as an unfair subsidy for river shippers. Some politicians agreed, insisting that a toll be charged for using the new lock. Congress had expressly prohibited tolls. Lawsuits proliferated. Hearings were held.

To resolve the fight, Congress set up an arrangement it hoped would satisfy all parties. When the Corps built something, it would rebuild landscapes commensurate with the ones it destroyed, and barge companies would pay half the cost of new navigation projects and major rehabilitations. An Inland Waterways Trust Fund would provide the industry's cost-share by collecting 20 cents for every gallon of fuel burned on designated inland waterways. The federal Highway Trust Fund did something similar with gas taxes.

To the subsidy averse, this still amounted to a huge giveaway, because the operation and maintenance of lock and dams remained a fully federal responsibility. Operations and maintenance are expensive and endless. With so many aging lock and dams and so few new ones under construction, some experts estimated that the federal government still paid between 90 and 97 percent of costs on the waterways.

Steve Ellis, vice president of Taxpayers for Common Sense, a nonpartisan watchdog, provided a cogent critique of the arrangement in testimony before a House subcommittee. In addition to their exemption from upkeep costs, he said, private companies that take advantage of the waterways also enjoy an Inland Waterway Users Board, "a federally funded, federally staffed board of private industry that recommends how taxpayers' money should be spent. None of the other transportation systems has a taxpayer-funded advocate sitting at the table." Ellis continued:

> The atypical cost-sharing structure of the inland waterways creates costly, unintended, even bizarre consequences. Since users don't have to pay anything for maintenance, they are constant cheerleaders for new construction. There is absolutely no recognition of the maintenance costs associated with the inland waterway system. There is no market mechanism to suggest that times have changed and certain waterways should no longer be maintained. There are federally maintained waterways that see almost no traffic in a year yet the taxpayer is on the hook to maintain the system. . . . After the Chattahoochee River saw only a handful of barge tows one year, former Congressman [Tom] Tancredo (R-CO) opined that it would be cheaper to ship by limousine.

Flood control, the Corps' primary mission along with navigation, is funded in a completely different way. Most levees were originally built by local governments or individuals. Via a series of Flood Control Acts,

Congress eventually put the Corps in charge of about half the country's levees, but these projects still must have local sponsors, most often Levee Boards with the power to tax property owners who benefit from the levee and apply that money to its operations and maintenance. On Flood Control Act levees, like those along the Lower Mississippi, the local sponsor is usually in charge of right-of-way, dirt, and property acquisition, and the feds pay for construction. On levees where the government has less of an interest, construction costs are shared 80 percent federal, 20 percent local. A levee built today might have a 65–35 cost share, as long as the benefits outweigh the expenses. If not, the feds won't pay a thing. In every case, if the sponsor can't afford its share, the government will walk away. If a levee is breached and the locals haven't been maintaining it, the government won't rebuild.

In 2014, after Ellis's testimony, the river shipping industry renegotiated its share of Olmsted's construction cost. Shippers argued that they couldn't afford to pay half of the new $2.9 billion budget, plus they still wanted to fund other navigation projects, like the Kentucky Lock Addition. The government allowed the barge industry to reduce its share of Olmsted's remaining construction cost from 50 percent to 15 percent. To raise additional money, the industry volunteered at the same time to increase its fuel tax to 29 cents per gallon. Waterways Council, Inc., a river shipping lobbyist, has been pushing to reduce the industry's cost share permanently to 25 percent. It says this arrangement will save taxpayers money by allowing the Corps to build projects more efficiently, and the Council remains vigilantly opposed to paying for operations and maintenance.

The American economy could not exist without navigation on the Mississippi, Ohio, and Illinois Rivers. Anything that hampers movement on these waterways will send declining incomes and increased costs rippling across the country. These projects have delivered overwhelming benefits compared to their costs, and a lot of this infrastructure was built during the Great Depression for the added benefit of creating jobs. Most of the really worthwhile rivers were dammed or channelized before World War II. Yet politicians, lobbyists, and the Corps came to rely on those federal dollars and, after the war, they proceeded to dam and channelize obscure rivers with less traffic. "Build it and they will come," proclaimed the Corps. Sometimes the promised jobs and economic development came, more often they didn't.

Water-resource bills have long been synonymous with the pork-barrel. Fiscal conservatives and small-government politicians periodically derided the pork, but the lure of money to spend and concrete to pour proved overwhelming.

Waterway projects were, for decades, funded via earmarks, line items buried in bills that piped money into a particular lawmaker's district. Often they were favors traded between politicians, or rewards to loyal interests back home. Earmarks were good ways to get something built efficiently, without a lot of committees and controversy. The practice wasn't intrinsically bad, but it made it easy to build expensive and unnecessary projects. An outstanding example was the lengthy effort to channelize and dam the sandy and shallow Arkansas River, covered in a scathing 1963 *Life* magazine article. *Life* called the Arkansas River Navigation project, then nearing completion, "the most outrageous pork barrel project in U.S. history." It was championed by Senator Robert S. Kerr, a Democrat and former Oklahoma governor who had become known as "King of the Senate." As *Life* explained:

> The pork barrel works best when those with an interest in it help one another. As chairman of the Senate's Rivers and Harbors subcommittee and acting chairman of the Public Works committee, Kerr was in the right position to put the principle to work. And he was ruthless enough to enforce it. Any other lawmaker yearning for a project back home automatically fell into Kerr's debt, and his associates remember that he was a hard-boiled creditor.

Three times, Kerr outmaneuvered President Eisenhower, who sought to defund Arkansas River Navigation. According to *Life*, Kerr also helped the Corps justify the project's benefits despite the rising costs by shepherding a law that changed the amount of time a project had to earn back the money spent on it. When accorded one hundred years, instead of fifty, Arkansas River Navigation and other dubious ventures had more time to accrue benefits against the same costs, and easily won congressional authorization.

Since 1970, when the build-out of the Arkansas River was completed, towboats have been navigating the sandy, shallow river from its mouth at the Mississippi all the way to Catoosa, Oklahoma, 445 miles away. The project cost $1.2 billion and took almost thirty years to finish. Its official name is the McClellan-Kerr Arkansas River Navigation System, after Kerr and Senator John L. McClellan, a Democrat from Arkansas.

Then there's the J. Bennett Johnston Waterway, 236 miles of the Red River, dammed and channelized from Shreveport, Louisiana, to the Red's confluence with the Mississippi. Completed in 1994 for $1.9 billion, its champion was Senator J. Bennett Johnston, Democrat of Louisiana, chairman of the Senate Energy and Natural Resources Committee, and later a high-powered lobbyist. In the *Washington Post*, Michael Grunwald described the dearth of barges on the Red, and wrote of Johnston:

> He knows the Red River project has always had skeptics. Five con-secutive presidential administrations balked at funding it. In 1980, an environmental coalition said it had "one of the most flimsy cost justifications of any Corps of Engineers project." But Johnston always managed to ram it into the budget anyway. And even though only a few hundred of the expected 15,000 jobs have materialized so far, he's glad he did.

Grunwald portrayed Johnston as an American archetype who believed that human dominance over nature was "a good" in itself. Johnston, Grunwald wrote, "still exudes contempt for 'cost-benefit analyses' and similar wonkery. He knows in his heart that taming the unmanageable Red was a godsend for Louisiana, even if the paltry barge statistics don't show it yet. Numbers, he says, can be misleading. Beauty, he says, tells the truth. 'We created a wonderful lake a couple hundred miles long!' Johnston says. 'That's not in the cost-benefit ratio, you know, but it should count for something. How do you weigh the advantages of a beautiful river?'"

Earmark spending reached $16 billion in 2010, and the following year the practice was banned. "Earmark" had become a dirty word, a synonym for bridges to nowhere and special interests. Since 1936, the Corps had weighed costs against benefits when studying a project, arriving at a ratio, or score; the higher the number, the more a project was in the public interest. The cost-benefit score seems to have filled the vacuum left by earmarks. The Corps maintained that its process was strictly objective, weighing, for example, tonnage shipped through a lock against the cost of highway shipping, and measuring that against the cost of the lock's construction. The corrupt earmark days were over, was the message; economics was a science, and data couldn't play favorites.

In 1998, a group of Corps economists completed an analysis of a proposal to extend five locks on the Upper Mississippi and two on the Illinois River. They concluded that billions of dollars' worth of new construction could not be justified. Small nonstructural changes, like a modified waiting and assist boat program, were all that was needed to speed long tows through short locks. Senior Corps officials rejected the findings and commissioned a new analysis. The original study's managing economist, Donald Sweeney II, filed a whistleblower complaint with the Office of Special Counsel, accusing the Corps of deliberately falsifying cost-benefit data in the new analysis in order to arrive at a predetermined conclusion, to wit: Build the new locks and build them fast. The army's Office of the Inspector General investigated Sweeney's complaint, which included a trove of internal emails—"If the demand curves, traffic growth projects and associated variables . . . do not capture the need for navigation improvements, then we have to figure out some other way to do it." The inspector general's report vindicated Sweeney and condemned the Corps. The agency was on a mission to grow, the report found, and in order to grow it needed to build, regardless of benefits and costs. The report concluded that "an institutional bias for large-scale construction projects may exist throughout the Corps."

The National Research Council (NRC), part of the National Academy of Sciences often tasked with providing nonpartisan advice, was called in to find out—once and for all—whether those seven locks needed to be upgraded. The 2001 NRC report determined that "shippers and tow operators bear needlessly high costs because there is no traffic management system. Rather than waiting for a decade for relief from the congestion by extending the locks, shippers and towboat operators could enjoy immediate improvements through better traffic management." The NRC identified several ways to make locks more efficient that didn't involve building anything new, like designing a better way to couple and uncouple barges so large tows could move through small locks faster. (To this day, barges are tied together with steel cables, dangerous and time-consuming to loosen and tighten.) The report's implicit question: Should taxpayers spend billions of dollars on new structures because towboat traffic isn't managed and barges are still laced together with wire? Trains run on a tight schedule, and train cars can be coupled and uncoupled quickly and with ease. Why not barges?

Arguing that improved barge scheduling and better couplings can solve the problem of aging locks, said Deb Calhoun, Waterways Council's senior vice president, is like telling someone struggling to drive across the country in a Model T to get a better map. If investments in the inland waterways continue to decline, the system will still work, it just won't work efficiently enough. According to a 2019 report commissioned by the U.S. Department of Agriculture, while U.S. waterborne transportation is getting slower and more expensive, Brazil—with China's help—is busy expanding its railroads, grain elevators, and export terminals. U.S. agriculture is losing its competitive edge. By 2045, if American lock chambers haven't been built bigger and made more reliable, Brazilian grain will dominate the world market. Farm ground in the American heartland will go out of production, farmers will make less money, small-town economies will suffer further, and, as freight moves off the river and onto roads and rails, congestion, fatalities, and carbon emissions will all increase.

The NRC report, however, cast doubt on the Corps'—or anyone's— ability to forecast freight movements decades into the future. The Corps' tonnage calculations, it said, often seemed to rise in tandem with the traffic needed to justify big projects.

"Figures lie, liars figure," Andy Schimpf liked to say. In its reauthorization request for Olmsted, the Corps told Congress that 113 million tons of cargo would transit Lock and Dam No. 52 annually by 2020. The forecast was based on the assumption that more and more coal would float down the Ohio from mines in the Appalachians. In fact, cheap natural gas and Obama-era regulations crippled the coal industry, and traffic through Olmsted's reach of the Ohio hovered around 85 million tons in 2018. In 1975, the Corps had predicted that 123 million tons would transit Mel Price annually by 1999. In reality, the 1999 tonnage was 77 million; by 2019, it had dropped closer to 50 million. Tonnage forecasts figure heavily into the Corps' cost-benefit calculator. If real costs were weighed against real tonnage, it's a fair guess that Olmsted might never have been built. The same might be true of Mel Price.

The Corps appears to have inflated tonnage forecasts and minimized budgets for years to get large-scale projects authorized. Yet cost-benefit scores have remained all-powerful. These scores can make or break an industry, force people from their homes, or sentence a town to flood.

"On a spreadsheet, you can have subjectivity all over the place, but if at the very bottom all of a sudden it has three decimals, it's considered completely objective scientific prioritization—and it's not," Schimpf said, as he descended a dark stairway inside one of Mel Price's massive concrete towers. "When I go home and tell my wife I need a new fishing boat, I can justify it on paper real easily."

Lock projects with low scores are still being approved. "Chicka-mauga, that's the one I can't figure out," said Schimpf. "Almost no tonnage, just one company that uses it." Chickamauga Lock and Dam, on the Tennessee River near Chattanooga, locks an average of one barge a day. Yet the government and the towing industry are spending almost $800 million to replace the 360-foot lock chamber with one that's 600 feet long. The alternative is no lock, because the old chamber's concrete is chemically expanding and will eventually fail. In this instance, taxpayers are essentially subsidizing pleasure boats, which outnumber commercial users here by about six to one. Barge traffic through Chickamauga has dropped more than 50 percent since 1999, and its cost-benefit score is 1.0. Projects with a score below 2.5 are generally not built, yet Chicka-mauga was reauthorized after blowing its initial budget and is on pace to be completed in 2023.

R. Mark Toy was the two-star general in charge of the Corps' Great Lakes and Ohio River Division, which includes Chickamauga. A good listener, sweet and accommodating, he had a knack for telling people what they wanted to hear without seeming to pander. Luther Helland described him as "fatherly." Toy sat under the tent at Olmsted after the crowds, politicians, and most of the media had left. The towing industry "is telling us what their objectives are," he said proudly, as if he had fulfilled a promise. Toy explained that the Corps had come up with a process called "remaining benefit-cost ratio," an exercise in creative economics to justify completing a project based on how much money had already been spent. Before Toy could elaborate, Semonite called him away. The Corps' top general was marshaling people for a photo op. "I don't want to be all squished in," Semonite yelled. "You guys, getting so organized, you're killing me."

Now that Olmsted was finished, the towing industry was focusing on Chickamauga, the $1.2 billion Kentucky Lock Addition, and three lock expansions on the Lower Monongahela River, a tributary of the Ohio, for $2.6 billion. After that, the industry wanted to enlarge three locks on the Upper Ohio, including Emsworth, which replaced Davis Island ($2.3 billion total). Next, it wanted to upgrade the seven locks on the Upper Mississippi and Illinois Rivers that were the subject of Donald Sweeney's whistleblower complaint ($2.8 billion). While it's true that tonnage shipped through each of these locks has declined in recent years, and the coal traffic that justified the original Monongahela and Ohio River locks is unlikely to rebound, several key industries rely on these waterways, a fact that's not captured in overall tonnage numbers. Royal Dutch Shell's ethane cracker plant, with 5,000 construction jobs and 600 permanent jobs, needs the Upper Ohio locks; two nuclear plants and the Oak Ridge National Laboratory rely on the Chickamauga lock for waterborne transportation; and navigation on the Monongahela is critical for U.S. Steel's Mon Valley Works. Grain and fertilizer volumes on the Upper Mississippi have the potential to increase, too.

Schimpf owns a baseball-related small business and is far from anti-capitalist, but he said, "I think the industry should be paying a lot more than they do, and I don't think the locks are in nearly as bad shape as they would have you believe." If they were paying more, or were paying for operations and maintenance, he argued, the industry would be a lot more choosy about what they championed, and about the real benefits and the real costs.

It comes down to this, in Schimpf's view: The Corps and the towing industry should not be in charge of deciding which projects to build.

It's a conflict of interest. "When we talk about all these things," he said, "who's at the table usually? The shipper-carriers and the Corps of Engineers—the two people who have the most to gain or the most to lose." Both parties benefit when the federal government pays the bills. When Olmsted took longer and cost more, who suffered? Not the Corps' project managers. Not the contractors. The politicians didn't mind as long as they had jobs and dollars to dole out. The towing industry and the big agriculture conglomerates got their spiffy new locks in the end. Everybody won, Schimpf said, "except for the taxpayer."

Two days after the Olmsted ribbon-cutting, lockmaster Shane Byassee was alone in his windowless office in a nondescript building on the bluff above Olmsted. He was taking control of the Ohio River today. For the first time, Olmsted's wickets were rising and it wasn't an exercise. Tomorrow its Tainter gates would drop and Olmsted would hold back the river. A whiteboard on Byassee's office wall had a list of things to do. Between "DET Cord" and "revise crane for traditional," someone had written "pretend everything is OK—bury head in sand."

Though it was slow going, the dam was rising without problems. Morale was fragile, however, following last week's broken wicket. Mike Braden, the former division chief, told Byassee that as the wicket was being lifted, it had levered against unyielding sediment and snapped its bolts. Divers had since cleared the trench, Braden said. It wouldn't happen again.

Almost everyone from the old dams was now working at least part-time at Olmsted. Keith Browning, No. 52's former work leader, was there when the wicket broke. It "made kind of a clank noise, loud enough that everybody looked around," he said. At Olmsted, Browning's title was master of the deck, in charge of the new wicket-lifter barge—the *James M. Keen*—built for Olmsted at a shipyard in Louisiana. The *Keen* had a low pilothouse and two huge red cranes, one over the house, and one on the far corner. The far crane was for lifting wickets. The deck of the *Keen* was covered with winches and cables and other large metal items, many of which Browning didn't need.

So far, only Browning and Zach Tirey had been trained to operate the wicket-lifter. Each of Olmsted's 140 wickets took twelve minutes to raise. That came to twenty-eight hours of dam raising—if everything went perfectly.

After it clanked, the broken wicket was lowered and raised again. "Come time to bring it back up, when it broke the water, the dern thing was limp, hanging from the end of the crane," said Browning. The crew called management and twenty minutes later twenty hands were on deck. Everyone wrote statements detailing what they saw, as if the deck of the *Keen* were a crime scene. They took pictures. Hours passed while the wicket dangled from the crane.

Officially, sediment took the blame. But the lock workers knew that the broken wicket was one of just twenty-eight secured, by design, to the river bottom with six bolts instead of eight. Browning and his crew member Jesse Hall were talking it over later, back in No. 52's pump house. If the designers were so sure sediment was the problem, Hall wondered, why didn't they fix it during the twenty-five years they spent designing the dam?

The whole project seemed to be the product of a fractured bureaucracy. Design was in competition with Operations. All their requests had to filter through the contractors. No one had asked those who would use the equipment what would work and what wouldn't. Hall and Browning ticked off what they perceived as a dozen mistakes, oversights, and useless features. No one, it seemed, had thought enough about the way sand moves on the riverbed. On dives to the bottom of the Ohio, Hall had lived through underwater sandstorms, unable to see the end of his arm with his helmet-mounted flashlight. He'd seen sand fill underwater holes and trenches in minutes. The scientists who'd built models of Olmsted did simulate sand with various materials, but apparently they never interviewed the people who walked on the riverbed. As Browning put it, "Something that gotta work on paper but doesn't work in real life."

The day Olmsted went from wannabe to dam, Browning was on the *Keen*'s deck, scrubbing algae off the face of a wet wicket with a long-handled brush. Hall was in the new wicket-lifter's pilothouse, paying out cables from the Kentucky shore. Tirey was in the big crane's cab, picking wickets.

He raised wicket number 83, then 84. They both stood, but was there a definitive thunk? He hooked 85 and was lifting it, when 84

tipped back and sank into the river with a splash. A second later, 83
sank, too. The crew on deck stood still, watching the wickets go under.
Their expressions said: "I told you so; this will never work." But it had
to. Thirty years of labor and almost $3 billion had been bet on this
moment. All that time and money hung at the end of Tirey's crane,
rested at his fingertips.

Tirey opened the cab door, climbed down to the deck, and disappeared onto the towboat, which pressed the *Keen* firmly against the dam.
In the towboat's galley, he got a plastic container of meatloaf out of
the fridge and put it in the microwave. He wasn't happy with the new
wicket-lifter. Despite its GPS, sonar, and whatnot, lifting wickets was
easier at the old dams. The new crane was so powerful, Tirey said, that
it was easy to damage a wicket. Knocking them down and dragging
them up the old way wouldn't work at Olmsted. Tirey had to be much
more cautious.

Browning took over in the wicket-lifter's cab. He retrieved 83 without incident, then he went for 84. Two flat screens flanked the wind-
shield. His chair had a keyboard tray, a mouse, and lots of buttons and
a joystick on each arm. The Ohio was too murky for underwater cam-
eras; the lifter didn't have any. Browning had to rely on a sonar image
appearing on the screen to his left. He dipped the crane's hook into the
river to search for 84's catching bar. He sighted the wicket's outline, a
yellow blur on a black ground. The catching bar shone bright yellow
above what looked like a hump of sediment on the river bottom.

The lower left corner of the screen displayed a video game–style
image of what he was looking for: 3D shapes in red, gray, and brown,
wickets in raised and lowered positions, a hook at the tip of the crane.
It was perfect: no debris, no sediment, no water.

A button on another screen turned red, signaling a lost GPS connection. Browning made a phone call. He waited. The button went
green again. Barely nudging the joysticks, he closed in on the catching
bar. Just then, the angle of the sonar shifted and the bar disappeared.
Browning hesitated, then dipped the hook. He had it. As he drew in
the wicket, the *Keen* heeled forward, and his focus turned to the screen
on his right where red and green digits indicated pounds of pull. As he
guided the joystick, the numbers had to stay green, below 18,000. Any

higher and he could rip the wicket out. Browning fingered the buttons. The numbers stayed green, and 84 surfaced into the sunlight.

But would it stay there? When this wicket fell beneath Tirey's fingers, a few minutes ago, sand and debris once again took the blame. Olmsted's wickets, like those at Nos. 52 and 53, rested against a prop, a long metal rod countering the force of the river. The prop needed to slide along and then set in a depression in the sill called a hurter. If the hurter was full of debris or sand, the prop wouldn't set and the wicket would fall back down. With the hurter invisible behind the wicket, the cab's computers offered no help.

Browning raised the wicket a little higher than the one beside it, released it, and waited for a thunk. No thunk. He tried lowering the wicket, but it sank with the hook. It had no spine, no strength to stand on its own. For an hour, Browning moved the wicket up and down, hoping the action of the prop dragging across the hurter would dig out whatever was in there. The spit bottle he kept close by slowly filled with tobacco juice.

On the deck, watching, the crew waited for a fleck of daylight between the catching bar and the hook, which might mean the wicket had found its footing. But it kept falling back. The men sat in scraps of shade, necks slathered with sunscreen, munching electrolyte ice pops. In the whine of the machinery, conversation was impossible below a yell.

The crane stopped moving, holding the wicket in mid-arc. The cab door opened and Browning climbed down. He stared at the wicket—his mystery, his adversary. He spat and walked away. It had been an hour and a half.

It was Tirey's turn again. In a few minutes he had 84 standing. He released it. It stood. He pushed back against its top with the knuckle of the hook, and the wicket's bottom flexed out as intended, stiff and solid. Tirey fished up 85. It thunked. So did 86 and 87.

Then Tirey stopped. Five wickets had been raised. Now it was time to wash the next five with an underwater nozzle. Washing had become standard protocol after last week's bolt breakage, but some lockmen thought the policy was bogus. "That one just set perfect, now we're going back to warshin!?" one said. He didn't think that sediment was the problem; moreover, catching and raising wickets was more than mechanical: It involved feel and art and luck that, once interrupted, might not return. Tirey had just proved it. When a guy like that found his groove, you didn't want to stop him.

The sun dipped toward the Mississippi, lighting up the eddies and upwells spiraling in series away from the standing wickets, the river a crinkled sheet of foil. Browning was in the cab, manipulating the washing nozzle at the end of the boom. The nozzle's spray was visible on the sonar as a golden extrusion, burbling down in crests, mixing with the billowing sediment. It was hard to tell what was happening. Hopefully he was preventing the next broken wicket, though Browning still wondered if the river bottom was to blame. "They handed us an unfinished product—and told everybody else it's finished," he said. "I don't know if that's the way it is with big business corporations. It seems kinda shady here, if you ask me."

PART II

Alluvial Empire

The land looked steamrolled, utterly flat. Yet Lester Goodin could see a subtle and consequential topography. "I probably have an eyeball like a micrometer," he said, driving his dinged red pickup across the Southeast Missouri floodplain, "like a captain of the sea develops a very close eye for the weather, the waves.

"We dropped off that little ridge," he said. The road may have dipped a bit; it was hard to tell. This was where the soil "goes from sand to gumbo," Goodin said. His "ridge" was probably an ancient bank of the Mississippi River. Meandering across its floodplain over millions of years, the river had abandoned this channel, but the sandy bank remained; so did the backwater swamp where fine clays settled and became "gumbo."

All the earth from here to the Gulf of Mexico began as river-borne sediment, carried in from anywhere within a V-shaped funnel of territory encompassing 41 percent of the coterminous United States and a bit of Canada. The river has built some fantastically fertile land. Goodin, who farms in Mississippi County, Missouri, on the northern border of the Alluvial Empire, often says that his rich, black soil is "the best that Ioway can send me." Alluvium is what a river deposits. The empires are of cotton, corn, and soybeans, grown in a roughly football-shaped swath of land stretching from Illinois to Louisiana. "In Mississippi County," said Goodin, "if you find a rock, it's because a human toted it there." On his farm, you'd have to dig down 200 feet to find rock. In New Orleans, it'd be six miles.

A seventy-four-year-old with a halo of white hair, Goodin is known around this stretch of the river as Mark Twain. He has played the part in a local production of the musical *Big River*, wearing a white suit, chewing a cigar, and swirling whiskey in a glass. He speaks in an orator's mellifluous bass and keeps handy a supply of Twainesque aphorisms, like: "A river expert is someone who has guessed right twice in a row, and a damned fool is a river expert on his third guess."

There aren't any dams on the Mississippi south of St. Louis. Life on the floodplain depends on levees, those long earthen walls that define the landscape of the Lower Mississippi Valley. In 1717, French colonists built America's first levee in front of New Orleans, the modern city's French Quarter. Soon French property owners were required by law to levee off their ground or forfeit it to the Crown, similar to Western home-steading requirements a century and a half later. The actual builders of America's first levee were likely enslaved Africans, or Native Americans. Enslaved people built many early levees, most of which began as local enterprises—farmers (or those they conscripted) working together to push up some dirt between their ground and the river.

More levees meant higher floods, because the same amount of water was standing up instead of spreading out. As less and less water was allowed to escape the river, levees up and down the valley had to be raised. In 1717, when the Mississippi was free to overflow almost every-where, a three-foot levee was enough to protect New Orleans. Today, the floodwall around that city is twenty-three-feet tall. Critics say that levees increase flooding. They do. But that's because Americans want to live and work in floodplains. If millions of people and trillions of dollars' worth of property could be moved, the levees could be torn down, and the flood crests would come down, too.

By 1900, both banks of the Mississippi had been leveed for a thou-sand miles. A twenty-foot-tall earthen levee protects Missouri's Missis-sippi County. Beginning in the village of Commerce, Missouri, it runs south, rising to a height of three stories near the Arkansas-Louisiana border, and then falling gradually until it goes to ground at Venice, Louisiana, a dozen miles from the Gulf.

Historically, much of the Mississippi Valley—Goodin's beloved flat land—was swamp. Starting in 1849, Congress passed several Swamp Acts incentivizing the clearing and draining of the great wetlands that once lined the river. The old cypress trees were cut down and an elabo-rate network of ditches was built to drain the land. The swamps became valuable farmland, worth defending from the river. Levees disrupted natural drainage, so pumps were often required to push the water up, over, and into the river. "Drainage in flat land is the name of the game, it overrides all else," Goodin liked to say. Once drained, an acre here will yield sixty bushels of soybeans or two hundred bushels of corn with little irrigation and less than an average dose of fertilizer. Goodin inherited a saying from his father: "That dirt is so good, you put a little salt on it, you could eat it."

As Goodin drove through fields of rice and soybeans, the levee appeared on the horizon. It was a deep shade of emerald. Two and a half million acres, including all the land in sight, depended on this dark, meandering line. The Commerce to Birds Point Levee, as it is called, protects more square miles than any other in the United States. If it were to breach, water would roll across the floodplain for 175 miles before it returned to the river. Goodin sits on the board of Levee District No. 3, which shares responsibility for Commerce to Birds Point with District No. 2. Boards govern almost all levees, reflecting the groups of farmers that raised the money and dug the earth to build the levees in the first place. Levee boards are political entities. Board members are elected, but votes are apportioned per acre, so a big landowner has more votes than a small one. In rural areas like Mississippi County, prominent citizens belong to the boards and their seats are passed down through generations. Goodin has sat on his board for twenty-five years; his father and uncle were on it, too.

About half the nation's levees are entirely private. The government doesn't oversee them, nor will it rebuild them if they fail. The other half are enrolled in some kind of federal program and are overseen by the U.S. Army Corps of Engineers. Depending on a levee's importance, the government may play a larger or smaller role in its upkeep, but every federal levee must have a local sponsor to pay for operations and maintenance. In-kind contributions—usually in the form of dirt—are also accepted. These may sound like modest responsibilities, but some sponsors are poor. Goodin's District No. 3 pays its share by selling crops raised on land it owns, but District No. 2—like most levee districts— collects a property tax based on acreage the levee protects (and not, by any means, on the entire two and a half million acres).

Members of both districts once touched on the subject of upkeep with a couple of visiting generals from the Army Corps who were also members of the Mississippi River Commission. At an abandoned gas station near Birds Point, the generals arrived late to meet a knot of farmers leaning against their pickups. The farmers wore jeans. The brass wore combat fatigues. Talk turned to the cost of levee improvements.

"Can I ask you what your annual budget is?" said a general from California, who was wondering how hard it was for the district to afford some upgrade or repair.

"We take in thirty-six hundred dollars," a man from District No. 2 told him. "That's what we take in each year." He allowed that the district made a little more by letting people cut hay on the levee.

Goodin could see the bales as he drove, scattered along the slope of the levee. For years, this section had been lower than it was supposed to be. The Corps wanted it raised, but the locals couldn't afford the dirt, so a crucial federal levee continued to deteriorate. Finally, Goodin helped broker a deal with a doctor who wanted a duck pond dug. The Corps came in and dug him a world-class duck pond and took the earth for the levee, which it hurried to raise and broaden. Goodin remembered watching the contractors work. "Did you know, building a levee, they shave down the ground and then cut down a trench?"

A levee is usually built with a three-to-one slope: At a height of 20 feet, it will be 120 feet wide plus the width of the crown, or top. Construction takes place during low water, usually summer and fall. The base is first anchored to the ground with a muck trench, a slot cut into the ground roughly corresponding to the levee's centerline. All vegetation is removed, then comes a linear mound built of earth or, if the district is poor—like many on the Missouri River—sand. High-end levees, like those on the Lower Mississippi, are built out of clay, or with a clay layer on the river side. Then the levee is grassed over and a gravel road is laid along its top. Only certain grasses are allowed: Bermuda on the mainline Mississippi, fescue in other places. Trees are prohibited. Their roots would rot away, creating inlets for water; trees toppled in storms leave craters where they stood. Digging varmints like feral hogs, badgers, and groundhogs are shot on sight. Even a small fissure can weaken a levee enough to collapse it in a major flood. Seepage-prone levees are finished on the land side with a berm, a reinforcing strip of grass-covered earth.

During times of high water, teams of muddy-booted riders will patrol a levee twenty-four hours a day. They carry powerful flashlights, on the lookout not only for landslides, but for every levee's nemesis: the sand boil. A rising river applies immense pressure to its banks and bed. If this water encounters a layer of exposed sand, it can move laterally between the grains, force its way under a levee, and explode upward on the land side. Some sand boils are inconsequential, a safe way for the river to relieve pressure, but, left unchecked, the moving water may begin to erode soil from beneath the levee. Too much erosion and the levee will collapse.

"I've seen sand boils as big as a tennis court," said Goodin. In 1981 and 1982 there were floods over Memorial Day weekend. Goodin and his neighbors were fighting to protect twelve thousand acres that had no levee. Crops worth hundreds of thousands of dollars were in the ground; they didn't want to let the river have them. The farmers built an eight-foot levee in three days. It seemed like everyone who owned or worked that ground was out there with a tractor and a dirt scoop. "It's wonderful," Goodin said. "Kind of like in the old days, if you go to a barn raising—stressful, but a lot of fun." They beat the river by six inches the first year, by less than a foot the next. "A farmer will do anything to save his crop," said Goodin, "—anything."

For communities living behind levees, floods are the events of a lifetime. In wind and rain and through the night, people summon every resource to stave off disaster. Say, "in '93" to anyone on the Upper Mississippi and they'll immediately recall the flood. A woman in Iowa does business only with people who were there in a flood year, fighting alongside her and her husband. One Corps engineer teared up remembering the moment twenty-five years earlier when a levee was lost to a sand boil. A man who fought the same flood wrote a letter to his children before he died, decades later, saying, "Never put faith in your earthen levees."

Sand boils range from catastrophic mega boils to inconsequential pin boils. Rickey Mathews and his crew from the Fifth Louisiana Levee District have seen them all. In March of 2018 they were patrolling a stretch of Mississippi River levee in northern Louisiana, between Waterproof and Vidalia, when they saw reddish water in a swampy wooded area beyond the berm. From notes passed down through three generations of patrollers, they knew this was a historic trouble spot, known as the "Hog Hole" for the feral hogs frequently encountered in the slough.

Mathews suspected a sand boil, and called in two geotechnical engineers from Army Corps' Vicksburg District, one of whom was twenty-nine-year-old Porter Holliday. The two engineers waded in to investigate. Before they saw it, they heard it—a bubbling sound, like a pot of water on high heat. The boil was breaking the surface of the slough, spurting into the air. Beneath the upwelling water was a fourteen-foot-wide cone of yellow sand, a model volcano with sloping sides and a crater in the center. The mouth of the boil, nine inches across, was spitting up balls of clay and gravel. Beyond the tumult, the calm water of the slough was covered in red-orange foam from subterranean colonies of iron-oxidizing bacteria.

The first step, Mathews and the engineers agreed, was to block off a nearby culvert to raise the level of the slough. The weight of the slough's rising water, they hoped, would put enough pressure on the boil to prevent it from moving earth. But this boil was too big. Even with the culvert stopped up, the cone rose higher and continued to spit clay.

The levee was in danger of collapse. In 1927, when a levee burst at nearby Cabin Teele, water flowed west for sixty-five miles. A breach here could be similarly disastrous. Like many boils, the Hog Hole was at the head of an old river course. The Mississippi has moved around a lot, even in recorded time. The Corps once commissioned a map of the river's previous channels. Geologists found thirty of them and traced each in a different color. Goodin compared the map to a snake held by its tail. Sometimes, an abandoned channel will silt in at its top and bottom to form an oxbow lake. The boil Mathews and Holliday were wrestling was between one such lake and the Mississippi. As the river rose, the lake rose, too, a sign that they were still hydrologically connected, likely through a layer of sand, which conducts water like copper conducts electricity.

The flood fighters' best hope was to surround the boil with a ring of sandbags. If the ring could be built high enough and wide enough, the weight of the water collecting in it would press on the boil until it stopped eroding. (You don't want to shut off a boil completely; it'll just appear someplace else.) The Fifth Louisiana Levee District was short of bags, sand, and manpower, but for as long as Louisiana has existed, levee districts have conscripted labor as needed; first enslaved people, later sharecroppers, and now prisoners. The levee district called Homeland Security, which called the local police jury. The Concordia Parish Correctional Facility and the East Carroll Parish Detention Center were already supplied with sand, bags, and filling machines. The levee district sent them a truckload of extra sand, and the trustees went to work.

At the levee, in a downpour, thirty-five prisoners formed a bag brigade, yelling "brick" down the line if a sandbag was particularly heavy. Mathews tried to explain to the prisoners—who seemed to be having too much fun—that if the boil weren't contained, they might wake up underwater. Slowly, the ring rose: eight bags wide at the base, narrowing to two at chest height. It took four days and 8,500 sandbags to tame the boil.

The districts had fought the same fight in 2011, except then they'd had twelve bad boils to contend with. They used close to 800,000 sandbags that year, on boils with names like Ice Box Hole, Diamond Island, Black Hawk—and Cabin Teele. From their notes, they knew that Louisianans had been fighting boils at these same spots for more than a hundred years.

Holliday had a technical and voracious mind and seemed to remember everything he'd learned about seepage and piping in his classes at Louisiana Tech, but it was another thing to see the river in action. On the fifth day of the 2018 flood fight, he paid the boil a visit, as he had every day since it was discovered. Twenty feet above him, the Mississippi was flowing by at a million cubic feet a second, a hundred times the volume of Niagara Falls. In the slough, the sound of flowing water was steady, like a backyard fountain. The boil was still alive, billows of fine sand rising from its throat. It was submerged in a few feet of clear water now, cut through with beams of sunlight, surrounded by lush bayou vegetation. Holliday, in waders, stepped off the ring of sandbags and walked out to the cone. He reached into its mouth and brought up a handful of fine, unthreatening sand. Earlier, the boil had been spewing coarse gravel. It took a lot of power to move such heavy material. Holliday figured the boil had transported about twenty cubic yards of it—"a nice dump truck load," said Mathews—before it was brought under control.

The water was warm, heated by the friction of moving subterranean sand. A spillway had been built into the sandbag ring, letting a steady waterfall gurgle out. Holliday was content to let it flow. "It looks bad, and a lot of the general public will get worried—'This field's always been dry and now there's a foot of water'—but as long as material's not coming out, it's fine. It's safely relieving the pressure," he said.

Holliday grew up in the Mississippi Delta, north of Vicksburg, in what's called the Yazoo Backwater, where "boil" is pronounced "bowl." His parents' land flooded in 1973, before the levee meant to protect it was completed. Their yard still goes under when the backwater gets high. For Holliday, flood protection is personal. "A lot of people couldn't live where they do without our levees," he said. "A lot of people don't see all the work that goes into it."

The sun was low, casting golden-hour light onto his profile as Lester Goodin drove out of the woods into fields of soybeans. "God, I love this land," he said.

Goodin parked near his property line and got out. This was Thompson Bend, where he owned 665 acres of alluvial bottomland, stretching to tree lines in three directions. He waded into the soybeans, knelt, and crumbled a handful of soil. In the jungle of beans, he discovered a shriveled weed splayed around a stump of stalk. "Look at that!" Goodin crowed. "I like them to suffer." It was a pigweed, *Amaranthus palmeri*, put to death by a new and potent herbicide called dicamba.

On the way here, Goodin had pointed out a field of soybeans engineered to resist death by dicamba. The field didn't appear to have a single surviving weed. "It's just perfect," he'd said. "Terrible farmers have beautiful crops, just like they did when Roundup Ready came out." The genius of Roundup was that it killed everything except the plants that were genetically modified to resist it, called "Roundup ready." Dicamba worked the same way. And yet, Goodin said, pointing to a virile stalk in a non-dicamba field, "Pigweed makes a million seeds, half a million seeds per plant. That is an evolutionary machine." All it takes, he said, is one seed that's slightly resistant. "Everything else will die, but it will survive and make a half million seeds." Pigweed had already developed a resistance to Roundup, once the deadliest of herbicides. That took a decade, at most. Goodin was impressed with the inventiveness and virility of the weed. The biotech companies would continue to feverishly update their herbicides in an endless arms race with nature, but, he concluded, "In the long run, I think pigweed may be winning."

Perhaps, but at that moment dicamba was the talk of every farm. The chemical was just too good. Under certain conditions of humidity and air pressure, it would form a vapor and drift through culverts and across fields, killing fruit trees, garden shrubs, and crops not bred to resist it. In Arkansas, a man was murdered in a dicamba dispute. Farmers had sued each other and Monsanto, dicamba's maker. Several states tightened regulations, and farmers began spraying a new formulation, supposedly less volatile. But dicamba-resistant seeds, also made by Monsanto, continued to dominate the market, just as Roundup Ready seeds, another Monsanto invention, had done a generation earlier.

(Bayer, a German pharmaceutical company, purchased Monsanto while a series of lawsuits alleging that Roundup caused cancer were coming to trial. In several cases that were decided, judges found that Roundup had caused the plaintiff's non-Hodgkin's lymphoma, a kind of blood cancer. The value of Bayer—the entire company—fell to just slightly more than it had paid for Monsanto a few years before. In June 2020, Bayer agreed to settle about ninety-five thousand Roundup suits for $10.9 billion. The company also announced a separate $400 million dicamba settlement.)

American farmers outside California generally don't produce produce. They grow raw material. Crops, like crude oil, are priced and traded on global markets. A poor harvest in Brazil will affect soybean prices in Missouri. Americans eat commodity crops only in processed forms. Almost no corn becomes corn on the cob. Almost no soybeans become edamame. Half of American soybeans are crushed into oil or meal for domestic livestock; the other half are shipped overseas. Most corn stays in the United States and is either refined into ethanol, or cracked, rolled, or ground for livestock. There are some very big U.S. farmers, but most work a few thousand acres for themselves, or for a family-owned LLC.

They are relatively small businesspeople sandwiched between corporate titans. They buy seeds manufactured by big pharmaceutical and chemical companies—DowDuPont, Bayer, Monsanto. And they sell the grain raised from these seeds to agribusiness giants like ADM, Bunge, and Cargill, which may own the entire commodity crop supply chain, from grain elevators in St. Louis to terminals in Yangjiang.

The vast levee network protects the farmland where grain is grown, and grain is among the biggest commodities moving on the Mississippi and Ohio Rivers, where an equally vast system of locks, dams, and dikes makes transportation cheap. Up to 60 percent of U.S. farm exports move by water, and most of this is offloaded from river barges onto oceangoing ships at terminals below Baton Rouge.

Because grain at those terminals originates from as far away as Minnesota, any delay or malfunction in the inland navigation system—a broken lock gate or a shallow spot in the channel—will affect the freight rate, and that will affect what grain elevators pay farmers. The difference between the price at the Chicago Board of Trade (where U.S. commodities are traded) and what an elevator pays is called "the basis." Various local and global events can affect the basis, but mostly it reflects the cost of shipping. If transportation is unreliable or slow, the basis goes negative; if elevators need more beans for export, the basis can be positive. Modern farmers have basis apps on their smart phones; they negotiate futures and options (puts and calls) that would confuse anyone who doesn't work on Wall Street. Farmers might contract in the spring to deliver soybeans in January at a price 20 percent above the current basis. And, if they store their soybeans in their cylindrical silver bins, they can wait out market slumps for at least a year—if they can put off their loan payments.

Stepping back from the annihilated pigweed, Goodin gently touched a purple-and-white soybean flower, pendent below the plant's broad leaves. In two months, the leaves would fall, the beans inside the plant's pods would harden, and the harvest would begin. Goodin had retired from farming after he tried to lift a fifty-pound bag of ammonium sulfate and woke up flat on his back. He rents his land now to a friend, but he still markets his share of the crop and checks on Thompson Bend every chance he gets. He's been told that he loves his dirt too much.

In October, combines will take off across the fields, throwing off plumes of chaff and dust. A combine, the size of a small house, can cost half a million dollars. Like a jet on autopilot, it steers itself. The operator, sealed inside an air-conditioned cab gliding ten feet over the field, watches an array of screens where real-time stats, like moisture level and bushel-per-acre, flash past in multicolored bar graphs. The data can be saved onto a thumb drive and analyzed with special software to help farmers determine which seed varieties work best in what soil types, and with what fertilizers, herbicides, insecticides, and fungicides. In the spring, a soybean planter will mete out 160,000 seeds per acre in perfect rows, "like they've been sewed with a Singer sewing machine," said Goodin. When it's time to apply herbicide, a 120-foot-wide, GPS-guided spray rig will turn off and on automatically, so not a single plant is skipped or double sprayed.

"A farmer is obsessed with controlling the things he can control," said Goodin, "because so many of the most important things he can't control."

As labor costs rose during the second half of the twentieth century, technology allowed farmers to work more land with fewer people. Small farmers either sold out or bought up their neighbors and got big. A farmer with fewer than a thousand acres couldn't afford to buy a combine, a spray rig, or a tractor anymore, said Goodin. "To justify the equipment you've got to have a lot more acres." Successful farmers joined the middle class, or better. Ex-farmhands and ex-sharecroppers joined the rural poor, thanks in large measure to John Deere, Case IH (International Harvester), and New Holland, makers of ever more efficient farm equipment. As Goodin saw it, in Mississippi County and in much of rural America today, "Agriculture is *the* industry and agriculture no longer requires the people." Fifty percent of Americans worked in agriculture in 1870. By the year 2000, it was less than 5 percent. Those workers sustained local businesses, governments, school districts, and county road budgets, all of which suffered when high-priced equipment took away jobs.

Goodin lives in an old Arts and Crafts house near downtown Cape Girardeau, Missouri. The home is comfortably cluttered. His crammed bookshelves reflect the time he spent pursuing a master's degree in history at the University of Chicago. He had "aimed to teach" but didn't complete his thesis. After college, he joined the Peace Corps and went to Nepal—"two days' walk from the nearest wheel." After a year and a half in the Peace Corps and two years as a Peace Corps bureaucrat in Washington, DC, he left for graduate school. In 1972, he returned to Missouri and took up farming.

William Faulkner, Goodin said, described "the environment I grew up in." Faulkner's territory was north-central Mississippi, while Goodin's was the Missouri Bootheel, the block of land that drops below the state's otherwise straight southern border. The Goodins were Bootheel patricians, farmers going back five generations: "Nobody's farmed some of our land except the Indians and a Goodin." They ran one of the bigger farming operations in the region, spanning roughly five thousand acres. Farmers usually own some fields and lease others, paying the owners a percentage. At one time the Goodins farmed for thirteen different entities. Lester grew up on the edge of Charleston, the county seat, in a Victorian mansion built by the family in 1899 at the corner of Main Street and Goodin Avenue. From his room in the second-floor turret, young Lester could lie in bed at night and see the spotlights of boats on the Mississippi. He'd say his prayers: "Now I lay me down to sleep, I pray the Lord my soul to keep . . ." And at the end, he'd add, "I want to be a riverboat pilot." Though Goodin did play Mark Twain on stage, he never realized Twain's boyhood ambition. Goodin's life was shaped by what he saw through his window during the day: the leafy avenues of the town in one direction, the low horizon of the fields in the other. Goodin became an educated urbanite and a passionate farmer. "The Mississippi runs in my veins," he said. "It is a profound part of who I am. Like Faulkner says, You don't own the land, the land owns you."

Though he was from the hills, Faulkner spent a good deal of time in the flatlands of the Mississippi Delta. His friend and literary agent, Ben Wasson, was a flatlander from Greenville. According to Wasson, the two men were watching the sunset on the levee near Greenville in 1924 when Faulkner exclaimed, "Listen! You know what the Old Man is saying. 'They're not going to tame me.' And nobody ever will." In his novel *The Wild Palms*, Faulkner cuts back and forth between two stories. One of them, called "Old Man," is set during the Great Mississippi Flood of 1927, the worst flood in American history, and the one that changed forever the way the country thought about river engineering.

Faulkner doesn't dramatize the deluge in "Old Man." To sense that drama, Goodin recommends *Rising Tide* by John M. Barry, the definitive history of the 1927 flood. Up and down the valley, *Rising Tide* is esteemed above every other book, except maybe Mark Twain's. Barry quoted a diary kept by Henry Waring Ball, a member of Greenville's upper-class planter society, who described the weather almost daily. Starting on March 7, 1927, and ending on April 8, his entries read, in part: "Rainy"; "Pouring rain almost constantly for 24 hours"; "Rain almost all night"; "After a very stormy day yesterday it began to pour torrents about sunset . . . unrelenting flood came down for four hrs. I don't believe I ever saw so much rain"; "A tremendous storm of rain, thunder and lightning last night, followed by a tearing wind all night"; "Torrent of rain last night"; "Bad. Cold rain"; "Still cold and showery"; "Very dark and rainy"; "Too dark and rainy to do anything"; "Violent storm all night. Torrential rains, thunder, lightning, high winds"; "Rain last night of course"; "I have seldom seen a more incessant and heavy downpour until the present moment . . . the water is now at the top of the levee."

Beginning in December of the previous year, a pattern of fronts and storms, trending southwest to northeast, had pounded a swath of the Mississippi River Basin from Texas to Pennsylvania. The Ohio and Arkansas Rivers' drainages were hit particularly hard. From December 18 to April 29, 1927, Johnsonville, Tennessee, on the Tennessee River, received 43 inches of rain. Over the same period, Nashville got 34. More than half the rain gauges in the Arkansas Basin registered 30 or more. Monroe, Louisiana, on the Red River got almost 45. Memphis: 40. Greenville: 43. New Orleans endured 14.96 inches in eighteen hours.

The rain that fell in the first half of April transformed a big flood into a super-flood. Substantial levees lined the entire river, but they weren't built for a flood like 1927's. Nothing was.

The first government levee to breach was at Dorena, a hamlet in the Missouri Bootheel, not far from where the Goodin family farmed, and not far below the confluence of the Mississippi and Ohio Rivers at Cairo, Illinois. There, the two biggest rivers in the continental United States slam into each other, creating one of the most flood-prone and hydrologically complicated areas in the country. The Ohio almost always carries more water, and if it is high, it can act like a dam, backing up the Mississippi for dozens of miles. If the Mississippi is high and the Ohio low, the latter will run backward as far away as Lock and Dam No. 52 (which was completed two years after the Great Flood, and where those rare, current-less hours were used to repair wickets).

When floodwater meets a levee, the water has to go someplace else. When too much water is trapped between levees, it rises over the top or tunnels under. Various engineers had long argued for allowing the Mississippi to escape its levees in a controlled way, releasing just enough water to reduce the crest of a flood and take pressure off the surrounding area. The feasibility of this plan had been demonstrated in 1922, when a levee breach at Poydras, Louisiana, lowered the crest at New Orleans. But those in favor of such outlets failed to convince the right people. In 1927, the "levees only" policy remained in effect. The valley had no spillways and no reservoirs, just miles and miles of levees, and the river had nowhere to go but up. And out.

One hundred seventy-five thousand acres were inundated when the levee gave way at Dorena. But without the outlet created by that breach, or crevasse, the flood's crest may well have been higher. Writing in the *Monthly Weather Review* of December 1927, government meteorologist A. J. Henry stated unequivocally that "if the Dorena crevasse had not occurred, the crest stage at Cairo would have been 57.7 to 58 feet." In fact, four days after the Dorena breach, the river at Cairo crested at 56.4. Water was spreading onto the floodplain, not standing up between the levees.

Farther downstream there was no such relief. Levees remained under threat, particularly to the north of Greenville, Mississippi, where the Arkansas River pours into the master stream. Dozens of rivers ran high that spring, but the major contributors to the super-flood were the Ohio and the Arkansas. Levee breaches often occur below confluences, where two clashing rivers create eddies and turbulence and, of course, high water.

Everyone in Greenville knew the river was rising. For weeks, the National Guard, local sheriffs, and planters had been mustering workers to bolster the levees. The Mississippi Delta, also called the Yazoo Delta, is the delta of the Delta Blues. Before the Civil War, fortunes were made on the backs of enslaved people who worked its sprawling plantations. After the war, they subsisted in only slightly freer conditions as sharecroppers. In the fight to hold back the flood of 1927, whites forced many Black sharecroppers to work without pay, sometimes at gunpoint, heaving sandbags and hammering planks to strengthen the levee's slopes and crowns.

Black musicians testified to this cruel treatment in some of the first blues records. A year after the flood, Lonnie Johnson released "Broken Levee Blues."

Police say work, fight, or go to jail, I say I ain't totin' no sack.
And I ain't buildin' no levee, the planks is on the ground and I ain't
   drivin' no nails.

Five days after the Dorena breach, the river trickled over the top of the levee at Mounds Landing, a dozen miles north of Greenville. Within half an hour, fifteen hundred people swarmed over the spot, tossing sandbags into the growing breach. Despite their efforts, the trickle grew to a torrent; the earthen levee began to quiver, then appeared to boil.

"The roar of the crevasse drowned all sound. It carried up and down the river for miles, carried inland for miles. It roared like some great wild beast proclaiming its dominance. Men more miles away felt the levee vibrate under their feet and feared for their own lives," wrote Barry. "The crevasse was immense. Giant billows rose to the tops of tall trees, crushing them, while the force of the current gouged out the earth. Quickly the crevasse widened, until a wall of water three-quarters of a mile across and more than 100 feet high . . . raged onto the Delta."

The Mounds Landing crevasse inundated an area fifty miles wide and one hundred miles long with up to twenty feet of water. The whole Delta—the triangle of plantations and sluggish streams stretching from just south of Memphis, across to Greenwood, and following the Yazoo River to the bluffs above Vicksburg—was overcome. Hundreds were swept away and drowned when the water rushed in. Others climbed to their attics, hacked holes in their roofs, and waited for rescue. The levee itself was the only dry ground for miles around. Almost all of the Delta's 185,459 inhabitants were forced from their homes. Thirty thousand would leave for good. Tent cities and refugee camps proliferated. Violence and hunger were rampant. It was the Great Mississippi Flood that led some Black sharecroppers to join the Great Migration, moving north and carrying the blues with them.

Faulkner's "Old Man" describes the journey of two fictional prisoners who are delivered from a penal farm to the levee and conscripted to assist the rescue efforts. A gloomy post-breach stillness had set in. Approaching the flood zone, the prison truck first drove along an elevated road. "There lay a flat still sheet of brown water which extended into the fields beyond the pits, raveled out into long motionless shreds in the bottom of the plow furrows and gleaming faintly in the gray light like the bars of a prone and enormous grating." Then the furrows disappeared under water, then the road, and then everything else. "A single, perfectly flat and motionless steel-colored sheet in which the telephone poles and the straight hedgerows which marked section lines seemed to be fixed and rigid as if set in concrete . . . It looked, not innocent, but bland. . . . It looked as if you could walk on it. It looked so still that they did not realize it possessed motion until they came to the first bridge."

In an article published by the University of Mississippi, Phillip Gordon wrote about Faulkner's relationship to the Delta and the river:

> In "Old Man," Faulkner likens the Mississippi River to an old man going on a debauch and, in his drunken debauchery, spilling out of his confinement and reminding those who live in his shadow that they need always remember their precarious insignificance when the old man grows tired of control and predictability.

In 1927, the mad Old Man threw off his levee straitjacket and took back his ancient floodplain, recovering more than 16.8 million acres, almost the entire Alluvial Empire. Except for the tent cities and Red Cross shelters clustered in squalor on the levee tops, the landscape would have been familiar to Spanish explorer Hernando de Soto, who murdered and pillaged his way to the river in 1541 and died on its banks the following year. Garcilaso de la Vega, also known as El Inca, the son of a conquistador and an Incan noblewoman, published an account of the expedition in 1605. Garcilaso wasn't there but claimed to have interviewed several survivors. Whether his book, *La Florida del Ynca*, is literature or history can be debated, but his description of a flood on the Lower Mississippi is unequivocal:

> Which in the beginning overflowed the wide level ground between the river and cliffs; then little by little it rose to the top of the cliffs. Soon it began to flow over the fields in an immense flood, and as the land was level without any hills there was nothing to stop the inundation. . . . It was a beautiful thing to look upon the sea where there had been fields, for on each side of the river the water extended over twenty leagues of land, and all this area was navigated by canoes, and nothing was seen but the top of the tallest trees.

Centuries before anyone would dream of controlling such a mighty natural force, Garcilaso was able to view the flood with an impartiality that was lost on later Europeans who sought to rein in the river's excesses for their own benefit.

Bob Criss, a hydrology professor at Washington University in St. Louis, wants people to remember that the natural Mississippi behaves as Garcilaso described it. The chaos and destruction wreaked by crevasses like Mounds Landing were not natural. They were man-made. Natural flooding is a "gentle process," in Criss's view. "A lot of the stuff attributed to raging rivers and flooding are actual consequences of the failure of our structures," he explained. Given an entire floodplain, a river would flow gently across the countryside, rising barely a foot a day. Rage doesn't come naturally to a river. It's when a river breaks free of the man-made structures that confine it, when it's forced through a nozzle-like levee breach, that it scours away soil and flattens buildings. Criss, a gadfly often quoted out of context, is no friend of farmers or the Corps of Engineers, yet he might find an ally in Lester Goodin.

Though Goodin's livelihood depends on levees and his life has been devoted to flood protection, he agrees with Criss that some places are better left unprotected. "If you can't build a levee that's not going to fail, don't build one," Goodin said. "Nothing's worse than a failed levee." Thompson Bend, the land Goodin may love too much, is levee-less.

The name Thompson Bend originally referred to the shape of the river as it hooked around this lobe of land. The distance around the bend is fifteen miles, while the distance across the bend's narrow "neck" is only a mile and a quarter. A stream flowing down a shallow grade—whether it's rain on a driveway or the Nile approaching the Mediterranean—forms bends or meanders that exaggerate themselves until the crook of the stream becomes so narrow that it dissolves. When the river takes such a shortcut permanently, the neck is swallowed up and the bend becomes a back channel or oxbow lake. This process—common in the past, but antithetical to the engineered river of today—is called a cutoff.

In *Life on the Mississippi*, Mark Twain compared the shape of the river to an apple peel tossed over one's shoulder. He described the untamed Mississippi's "disposition to make prodigious jumps by cutting through narrow necks of land, and thus straightening and shortening itself." "The Mississippi between Cairo and New Orleans was twelve hundred and fifteen miles long one hundred and seventy-six years ago," Twain wrote, and "its length is only nine hundred and seventy-three miles at present." That was in 1883. The distance between those same cities has since been reduced to eight hundred and fifty-six miles. Twain joked that a scientist, looking at such figures, would readily conclude that "seven hundred and forty-two years from now the Lower Mississippi will be only a mile and three-quarters long," with Cairo and New Orleans side by side.

During the Civil War, Ulysses S. Grant tried to take control of the Mississippi, strategically and hydrologically, by forcing thousands of enslaved people and Union soldiers to dig a canal behind Vicksburg. Located on the outside of a sharp bend, the city commanded a high bluff that bristled with Confederate guns. If the river could be persuaded to bypass Vicksburg, the Union could take the Mississippi without engaging the heavily fortified city. But Grant's canal failed. The Mississippi stayed put. After trying in vain to find a detour through various bayous, oxbow lakes, and lesser rivers, Grant lay siege to the city. "Vicksburg held out longer than any other important river-town, and saw warfare in all its phases, both land and water—the siege, the mine, the assault, the repulse, the bombardment, sickness, captivity, famine," wrote Twain, who briefly fought for the Confederacy. Eventually, Grant did conquer the town, the Mississippi, and the South. The siege of Vicksburg was remembered as an especially grueling episode in a grueling war.

Thirteen years later, the river formed Grant's cutoff naturally. As he had hoped, Vicksburg was "left out in the cold," as Twain put it, "a country town now." Its harbor began to silt up and soon only small boats could call. With the river went the economy. It wasn't until 1905 that the Corps found a solution to Vicksburg's dilemma: Reroute the Yazoo River to give the town a new waterfront. That river cooperated. The new mouth of the Yazoo had enough flow to scour out the sand and maintain Vicksburg's harbor.

The levee around Thompson Bend failed long ago and was not rebuilt. By the early 1980s, a cutoff was proceeding across the narrowest part of the bend, which Lester Goodin owns. The Mississippi in flood would rip across the neck, gouging huge holes in some fields, covering others with sand. Goodin might have lobbied the Corps or raised money locally to rebuild the levee. Instead, he planted trees—ash for strength, cottonwoods because they grow fast, and pecans to generate a little cash. Farmers who work the flatland usually cut down trees, along with anything else that stands in the path of their combines, as they turn terrain that once hosted frame houses and kitchen gardens and post-and-beam barns into a washboard of row crops. Goodin and his neighbors did the opposite. They took eighty acres out of production forever and turned it into forest. The trees grew until Goodin and his neighbors couldn't see the river anymore. The old levee was long since gone, and the Mississippi was free to come in.

"Think of a river as an energy system," Goodin said. When the river rose, the trees absorbed its energy. Water covered the land, but it wasn't focused through the nozzle of a breached levee. The Mississippi's powerful current and coarse sand stayed in the channel, while the excess water overflowed harmlessly, dropping fine dark silt onto Goodin's land as it had for millions of years. When the Great Flood of 1993 surged through the Missouri and Upper Mississippi Valleys, demolishing levees, tearing apart homes, and killing fifty people, Goodin's trees withstood the deluge. Some cracked, but many stood their ground, cutting the water's speed in half. Instead of scour holes, Thompson Bend was blessed with two inches of new soil.

The Corps bought land upstream from Goodin's and planted trees there, too. It would have cost more to stabilize the banks with rock. Eventually, it made the program official and planted trees along the bend's entire length. Goodin even coauthored a paper: "Vegetative Based Solution to Controlling Overbank Scour in the Mississippi River Flood-plain Mississippi County, Southeast, Missouri," published in the *Journal of the American Society of Civil Engineers*.

Thompson Bend resembled a moonscape in the 1980s, before the trees. It has since become a perfect parcel of farmland. Goodin pointed to a muddy swale at the edge of the bean field. That was the would-be cut-off, the potential course of the Mississippi. The man who rents the bend from Goodin usually plants his soybeans in May, after the water recedes. An acre of the flood-washed land regularly produces sixty bushels. This delectable soil has earned Southeast Missouri and Southern Illinois the nickname Little Egypt, for a fertility that rivals the Nile Valley's. It is no accident that Memphis, Thebes, and Cairo are Mississippi River towns, and that Karnak and Luxora are nearby.

The headquarters of Levee District No. 3 occupies a baby-blue cinder-block building in Wyatt, across the river and nine miles south of Cairo. Wyatt is less a town than a collection of buildings occupied by ex-farmhands and ex-sharecroppers. One day, after a board meeting, Goodin stopped at Wyatt's only store to buy a sandwich: a half-inch-thick, perfectly circular slab of baloney on white bread, with a thin prefab piece of cheese and "lots of mustard" for $3. Goodin ate with one hand while he drove. The extended cab of his truck was full of fine dust and dog hair. A crumpled pack of Camels sat on the dash, an unopened one on the radio. Outside of town, he opened the window to smoke, accelerating on the straightaways and curves he had known all his life. He passed the farm where he'd been goose hunting on the day his father died.

A. Vernon Goodin III had been standing at the top of a grain bin while it was being unloaded. He must have lost his footing and fallen into the soybeans, which were siphoning downward like quicksand.

People were cutting holes in the side of the bin when Goodin got there. He jumped into the rapidly emptying bin and reached blindly down into the beans. His hand met his father's forehead, Goodin said, "and it was cool."

Vernon was sixty years old; his son Lester was twenty-nine. "Pappy, pappy," Goodin said, shaking his head as he told the story. "Why in the hell did you do that?" Vernon would keep those around him safe, but he took chances when he was alone. The Goodin brothers couldn't afford to lose the beans that had spilled while they were trying to rescue Vernon. In that same flood-fighting spirit, scores of neighbors arrived to help them shovel up their crop and deliver it to the elevator that night. Then the brothers buried their father.

Goodin had been working on his graduate degree in Chicago. After Vernon died, his three brothers asked him to join the business. "To hell with graduate school, I'm coming home," he replied. The four brothers farmed together for twenty years. Goodin took on Vernon's former duties: choosing seed varieties and fertilizers, analyzing yields and soil types. Each fall, he drove the combine to judge the results. He handled the money, too. The farm had a $1.2 million line of credit, and they spent about that much each year to make a crop. Like most farmers, the brothers always owed someone money. Goodin could recall only one time—a six-day span in 1973—when they were completely debt-free.

Though he loved his father, Goodin always felt closer to his uncle, his mother's sister's husband, A. J. Drinkwater Jr., whom he called Grandpa Dee. "When I talk about my father, I talk about two people interchangeably," he said. Goodin often stayed in town with Dee and Aunt Grace, who were childless. When Goodin finally married, at forty, his wife, Janet, already had three children. The couple had one son together. His name is Dee. None of the kids ever wanted to farm and Goodin was glad of it. They all grew up in town. How could they make business decisions? More important, how could they stay safe?

"Farming can kill you in ways you've never dreamed," Goodin said. "My father was killed. My oldest brother lost a finger. My second brother, John, had a back operation, two hernia operations, and spent a month in a burn unit. Lee, my younger brother, had the tip of his thumb cut off. And I've had five back operations and three operations on my neck.

Just in the past three years, my lower back has been fused twice. The prognosis ain't very good. And when you do it twice, it really ain't. I'm real tickled to be alive."

Goodin's eldest son works for the Missouri Department of Transportation as an engineer. His daughter is a teacher. His third son runs a family-owned campground with more than a hundred campsites, hiking trails, and streams for canoeing. Dee, the youngest son, runs the campground's lodge. Janet, Goodin's wife, owns a preschool in Cape Girardeau with one hundred employees and helps manage the lodge, a two-hour drive away.

The family settled in Cape Girardeau to be closer to Janet's school. A college town, it is the biggest city between St. Louis and Memphis. To reach his beloved Thompson Bend, Goodin drove south on Interstate 55, descending the Benton Hills. The floodplain stretching out before his speeding pickup was so flat that it seemed to bend upward.

One hundred million years ago, the central United States drained west into present-day Arizona. The Appalachian-Ouachita mountain range stood as a solid wall between the interior of the continent and the Gulf of Mexico. The tectonic plate on which the mountains and indeed all of North America sat was slowly passing over the Bermuda hot spot, deep beneath the Earth's mantle. A different hot spot made—and is still making—the Hawaiian Islands, as the Pacific plate moves northwest. Similarly, where the hot spot underlay the Appalachian-Ouachita range about 80 million years ago, the heated rock bowed upward. Exposed to the wind and rain, the arch eroded and, as it cooled, it sank like a fallen cake. The sinking breached the mountain front, and much of the continent began to drain through the cleft. Seawater also rushed in, filling the sunken area. This new waterbody, reaching to the middle of the continent, became known as the Mississippi Embayment.

The powerful new river now pouring through the cleft dumped sediment into the embayment, and land arose on both sides of its channel. The shoreline moved southward as the embayment filled and became a floodplain. The river's path to the Gulf squiggled and zigzagged like a garden hose with the water on full blast. The sediment formed a

six-hundred-mile-long wedge, heavy enough to depress the continental shelf. Geologists compare this massive deposition of material to
an inverted mountain range, as deep as Mount Everest is tall. Oil and
gas—the decomposed bodies of tiny animals and plants, millions of
years older than the river—migrated up into the lithified sediments.
Salt flowed upward, too, forming bulbous sulfur-laced domes.

When its channel became too shallow and choked with silt, the
river—which the Anishinaabe people called *Misi-ziibi*—veered from its
bed to find a shorter, steeper route to the Gulf. In the last five thousand
years, the river has used six different outlets. The natural process of
delta building and delta abandonment created all of south Louisiana.
The river's recently abandoned courses still exist as minor waterways—
Bayou Cocodrie, Bayou Teche, Bayou Lafourche—snaking down from
their junction with the current channel. If not for man-made structures preserving the status quo, the Mississippi would have relocated
again sometime before 1970, following the course of the Atchafalaya
(Ah-cha-fuh-lie-yuh) River to the Gulf.

After finishing his baloney sandwich, Goodin drove north toward
the village of Commerce. The Benton Hills rose up like a wall ahead of
his pickup, and the plain abruptly gave way to steep inclines and tight
valleys reminiscent of Appalachia. In Scott City, Goodin turned east
toward the river. From the ridge top, the view opened up again and he
spotted something new: a fancy bar safe from floods atop a man-made
mound. He stopped in for a beer and sat outside to drink it. Across
the river, the columned courthouse of Thebes, Illinois, glowed in the
setting sun. Trains rattled past on the steel truss bridge downstream.
A barge tied up below the town carried equipment for excavating the
river bottom, which, right here, was solid rock.

Ten thousand years ago, this is where the Benton Hills were cut in
half by a huge ice-age flood, revealing a cross-section of layered limestone
called the Thebes Gap. Never losing an opportunity, the Mississippi
flowed through the gap. If the Army Corps of Engineers could travel back
in time to the end of the last ice age, it would surely try to prevent this
change of course. The gap created intractable problems for navigation
and flood control that the Corps has yet to solve. Water has backed up
behind it, flooding Cape Girardeau, and it has blasted through the gap,

bearing down on levees, including the crucial Commerce to Birds Point section. In low water, rock pinnacles would scrape the bottoms of barges, restricting the width and draft of tows, until the Corps had the pinnacles blasted away with huge hydraulic hammers and lots of explosives.

Fifty miles to the south, near the town of New Madrid, Missouri, lies the Reelfoot Rift, a weak point in the North American Plate where millions of years ago the continent nearly split apart. Among other deformations to the surrounding landscape, a huge block of rock gradually tilted southward and its high side became the Benton Hills. The continent ultimately held together—there is no ocean between Tennessee and Arkansas—but the weakened crust wasn't done moving.

During the winter of 1811–1812, at least three large earthquakes hit New Madrid (pronounced MAA-drid), shaking the alluvial soil like a rug. Eliza Bryan was then thirty-one years old and living in her mother's boarding house. The town had recently belonged to Spain, but would soon be part of America's new Missouri Territory. In a letter to a Methodist minister, Bryan described what became known as the New Madrid Earthquakes.

> On the 16th of December, 1811, about two o'clock, a.m., we were visited by a violent shock of an earthquake, accompanied by a very awful noise resembling loud but distant thunder, but more hoarse and vibrating, which was followed in a few minutes by the complete saturation of the atmosphere, with sulphurious vapor, causing total darkness. The screams of the affrighted inhabitants running to and fro, not knowing where to go, or what to do—the cries of the fowls and beasts of every species—the cracking of trees falling, and the roaring of the Mississippi—the current of which was retrograde for a few minutes, owing as is supposed, to an irruption in its bed—formed a scene truly horrible.

Tremors were felt in Boston; people were awakened in Washington, DC; the ground oscillated in Detroit. Near the epicenter, fissures opened hundreds of feet long. Topsoil liquefied, shooting water, sand, and lignite as high as treetops. The log houses of New Madrid toppled. Timothy Flint, who passed by a few years later, wrote that the graveyard, "with all its sleeping tenants, was precipitated into the bend of the stream." Islands disappeared. Lakes were formed in an hour, others vanished. The Mississippi sloshed side-to-side in tidal waves. Boats were crushed by its caving banks and launched up its frothing tributaries. Dead trees lodged in its bed were wrested loose and flung into the air.

The noise issuing from underground, variously described as musket, cannon, and artillery, lasted for days, sometimes accompanied by flashes of light. Large earthquakes, later estimated at above 7 on the Richter scale, continued at intervals well into 1812. In Tennessee, the Reelfoot River became the Reelfoot Lake, as a swath of ground sank and filled with water, drowning everything but the cypress trees. The last great quake, on February 7, 1812, destroyed what was left of New Madrid and toppled chimneys 160 miles away in St. Louis.

The New Madrid Seismic Zone remains active. A 2009 report commissioned by the Federal Emergency Management Agency found that a 7.7-magnitude earthquake there would cause "the highest economic losses due to a natural disaster in the United States." Very few structures in the Mississippi Valley today are earthquake-proof. According to the report, a big quake would cause "widespread and catastrophic physical damage" across eight states. Nearly 715,000 buildings would be damaged, 7.2 million people displaced, and 86,000 injured or killed. The report estimated that "direct economic losses for the eight states total nearly $300 billion, while indirect losses may be at least twice this amount."

Chris McLindon, president of the New Orleans Geological Society, has shown how several historic levee crevasses have occurred along fault lines where huge blocks of ancient sediment are sliding past one another. In McLindon's view, the seismological landscape underlying the country's critical infrastructure is not well understood. Testifying before the Mississippi River Commission in 2019, he begged for funding to create a geohazard atlas. The commissioners listened politely, took note of his requests, and said no more.

Lester Goodin likes to say that "every flood is as different from every other flood as one human being is from another." The specific origins of really big floods still mystify the experts, who do not have a definitive explanation for the flood of 1927, or for the floods of 2017, 2018, and 2019. Why did the water rise so quickly? Exactly where did it come from? Was climate change to blame? Or farmers? Or development? How about the Army Corps of Engineers?

Floods depend on rain—a lot of rain—but precipitation tells only part of the story. Different factors must converge in a precise and unlikely sequence to create the conditions for a super-flood.

Think of a river basin as a vessel for storing water. The Mississippi Basin is roughly triangular in shape, stretching from Montana to New York and covering most of the Midwest and mid-South, then funneling down to south-central Louisiana. Before white settlers intervened, the basin stored great quantities of water. But land that has been deforested, graded, and ditched to promote agriculture stores much less. Houses, factories, roads, and Dollar Generals with their impermeable roofs and miles of asphalt, store no water at all. The water that goes unabsorbed runs off into streams, lakes, and rivers. If more water accumulates in a river than it can contain within its banks, it floods. In this way, events months in advance will contribute to a big flood's height and duration.

The destruction and death of 1927 finally convinced the nation to take flood control seriously. First under Herbert Hoover, whose work with the Red Cross during the 1927 flood catapulted him into the national spotlight, and later under Franklin Roosevelt, the federal government would reroute, confine, and hold back the country's rivers.

Losses from the Great Flood approached $1 billion, a third of the federal budget at the time, a proportion roughly equivalent to $1.5 trillion today. Up to five hundred people died as a direct result of the flood while hundreds more died from starvation and seven hundred thousand lost their homes. The piecemeal levee system had been reduced to rubble, and competing state and local interests were ready to let Washington take over. Citizens who had patrolled their levees with rifles ready to shoot saboteurs from across the river—who knew, as they did, that a breach in Arkansas might save Mississippi—were now tired, wet, and hungry.

Immediately after the flood, Congress ordered the Mississippi River Commission to devise a master plan for managing the Lower Mississippi, from Cairo to the Gulf. Engineers in and out of government debated strategies. Congress decided to adopt a proposal from the chief of engineers, Major General Edgar Jadwin, and the 1928 Flood Control Act authorized and partially funded a version of Jadwin's plan. Ninety years later, the work known as the Mississippi River & Tributaries (MR&T) Project is still under construction. As of 2020, it had cost $16 billion.

Under the MR&T Project, levees were set farther back from the river's banks. The undeveloped no-man's-land between the riverbank and the new levees was called the "batture," from the French for "to beat." The batture gave the river more room to safely overflow, and the trees planted in it helped protect the levee from the river's erosive force. In exchange for giving land to the Corps for the setbacks and batture, the Mississippi Valley's residents gained the protection of levees that were wider, taller, and better engineered than ever.

The project included new reservoirs near the headwaters of many Mississippi tributaries to store runoff that would otherwise feed the river. Jadwin also proposed the construction of five outlets, breaks in the levees leading to sacrificial pieces of land known as spillways or floodways, where river waters could be stored or diverted. As the 1927 levee breach at Dorena had proved, when a river is free to overflow, its crest will fall. Similarly, Jadwin included four "backwater areas" in his plan, where farmland would be fronted by levees intentionally built lower than those protecting the rest of the valley. When the river ran very high, the less-populated backwater zones would flood, relieving pressure on the mainline levees. But the first areas to go under would be the floodways.

The simplest way to get rid of excess Mississippi River water is to send it down the Atchafalaya River, the master stream's largest distributary. The Atchafalaya branches off the Mississippi at a lonesome spot in east-central Louisiana, then cuts south through a vast swampland before emptying into the Gulf near Morgan City. The Atchafalaya has always taken some water from the Mississippi. To divert more in times of flood, the Corps built the Morganza and West Atchafalaya Floodways. At Morganza, Louisiana, where the two rivers are less than nine miles apart, a structure with 125 steel gates was built into the mainline levee in 1953. Behind it, a swath of pasture was framed by guide levees to direct floodwaters across the land and down into the east Atchafalaya swamps. The West Atchafalaya Floodway—which has never been used—was designed to pull water off the main stem Atchafalaya, diverting it onto farm ground, past several small ring-leveed towns, and down into wetlands west of the river.

It was also logical to build floodways where levees had breached in the past, since the ameliorative effects of these crevasses were already known. There had been eight breaches near Morganza in the 1800s. The old levee at Bonnet Carré Bend, about thirty miles upstream from New Orleans, had likewise breached four times before 1882. The bend brought the Mississippi to within six miles of Lake Pontchartrain, a large brackish estuary north of New Orleans. The site of the Bonnet Carré Crevasses became the Bonnet Carré Spillway. Completed in 1932, its gates can be opened to divert water into the lake before the Mississippi hits New Orleans.

The region's northernmost floodway was meant to take pressure off the Mississippi-Ohio confluence by mimicking the Dorena crevasse. It was called the Birds Point–New Madrid Floodway, after the two Missouri towns where the river would enter and exit. At Birds Point, water would leave the river and flood 133,000 marginally developed acres of Missouri's Bootheel. After the land had absorbed its fill, the water would re-enter the Mississippi just above New Madrid. By inundating the few crossroads towns of the floodway, the Corps hoped to protect the entire confluence area, especially the then-booming city of Cairo, Illinois.

Another floodway, mimicking the levee breach at Mounds Landing in the Yazoo Delta, was discussed but ruled out, ostensibly because the water would return to the Mississippi, though that fact hadn't disqualified Birds Point–New Madrid, three hundred miles to the north. There wasn't a city like Cairo to protect in or around the Yazoo, and across the river in Arkansas a floodway was already in the works. Instead, a series of reservoirs were built in the hills east of the Delta to control the Yazoo River's tributaries, and a backwater area was planned near its mouth.

The government never bought out the private land in the backwater zones. It may have felt that the overall protection these low-value acres gained from the project was worth an occasional flood. Floodways, by design, would be inundated more often, so the Corps agreed to compensate those landowners. By purchasing flowage easements, the Corps could flood the land—in perpetuity—whenever the river reached a certain height. By law, however, half the easements had to be in government hands before construction began. People in the valley wanted flood protection, and the science behind floodways had been thoroughly proven. Yet when it came to individual landowners, few were willing to sell.

The Bonnet Carré Spillway covered a small area, so the government bought it outright. Everywhere else, it needed those easements. Lawyers came knocking, checks in hand. The checks were small, and landowners knew that they had no real choice to opt out—the government could take their land and give it to the river by invoking eminent domain. Though Jadwin's opponents fought furiously, in Washington and in the courts, all but one of his floodways were completed.

The next big flood came down the Ohio in 1937. The government had the easements it needed, and the Birds Point–New Madrid Floodway was ready to go. While later floodways were equipped with mechanical gates, the Birds Point–New Madrid levee was originally intended to overtop. This "passive" operation took the human element out of a decision to use the floodway. But, within a few decades of constructing it, the Corps was pressured into raising the Birds Point–New Madrid levee to a height that would overtop only during the worst imaginable floods. The solution—since the government still wanted to operate the floodway at will—was to include "fuseplug" sections in the new levee that could be loaded with explosives and blown up.

The Goodin family owns acreage in Birds Point–New Madrid, but no one has ever found any record of an easement. The family might have refused to sell. It made no difference. The government blew up the fuseplugs as planned in January of 1937. As the icy water surged through the breach, Vernon Goodin fired up his motorboat. Later he told his son how he steered through the rain, sleet, and snow, rescuing people trapped on roofs. As he passed a frozen field, a forty-acre plate of ice suddenly heaved up and moved off downstream. Vernon watched that ice sheet slice slowly, inexorably, into trees, making "the damnedest racket." From that day on, every generation of Goodins fought to get the floodway deauthorized.

Even before the waters receded, the *Kansas City Star* sent Missouri-born artist Thomas Hart Benton to sketch the flood-devastated Bootheel. Benton wrote about the experience in his autobiography, *An Artist in America*. "Descriptions can give no sense of the dread realities of flood misery—the cold mud, the lost goods, the homeless animals, the dreary standing around of destitute people," Benton wrote. "The roads of the flood country were full of movers." He watched beds, stoves, whole chicken coops, and even pigs loaded onto wagons, trucks, and Model T Fords. "Lord knows where they were going." Benton's sketches of these scenes later became the painting *Spring on the Missouri*, though it was the Mississippi that had flooded that spring.

It was not until 1942 that the federal government acquired all the Birds Point–New Madrid flowage rights required by law. Eventually it purchased easements on almost 80 percent of the land at an average price of $17 per acre, but many farmers, especially in the early years, received only 50 cents an acre. The checks were cut and cashed, the money was spent, and for three-quarters of a century nothing happened. A generation of farmers died, and another was born without ever seeing the floodway operated. Those who lived and farmed there knew that their land was, in a way, not truly their own—it was vulnerable, expendable, sacrificial—but that knowledge faded with each bountiful harvest. The levees had been raised again after 1937 and the river seemed under control. People began to believe that the government would never flood their land on purpose.

Then, in January of 2011, it started to snow. It snowed more in February, then it started to rain. The snow melted, and it kept raining.

*   *   *

And raining.

Twan Robinson didn't take much notice of the weather that spring of 2011. April rain was routine, and working forty hours a week at a mental health clinic kept her busy. She was driving a client to the dentist when her older sister, Debra Robinson-Tarver, called her to say that the government was going to blow the levee.

"Lady, we got to get out, we got to get out now," Debra said, calling Twan by her nickname, "Lady." She had better get home and move her stuff.

Twan hung up and took a deep breath. Seeing her face, her client asked what had happened. "I got to go home. I got to go get my momma out of this," she said.

The sisters and their extended family lived in the Black community of Pinhook, Missouri, twenty or so buildings, home to maybe thirty people, two-thirds of the way down the Birds Point–New Madrid Floodway. The sisters' grandfather, Jim Robinson, had helped found Pinhook in the 1940s. About fifty years later, their father, Jim Robinson Jr., was interviewed by a white man named Will Sarvis for the Oral History Program of the State Historical Society of Missouri. Jim Sr. was a successful farmer who went from being a sharecropper in Arkansas, to renting ground in Tennessee—where he had sharecroppers of his own—to saving up enough to buy his own farm. A brother-in-law living in the Missouri Bootheel convinced him that land there was cheaper, and race relations were better than in Tennessee, where, Jim Jr. told Sarvis, "We couldn't talk to whites as long as me and you have talked without fear of reprisal of some sort. This was another reason why my daddy wanted to get away from Tennessee and come to Missouri. Because in Tennessee during that time we could not see the prospect of getting grown. And my daddy didn't bow to nobody, you know, and I'm similar to him."

Missouri was still heavily segregated, however. In the Pinhook area, Jim Jr. remembered, Black children went to school in wooden buildings with wood-burning stoves and secondhand books, while white children went to brick schools with coal stoves and new books. White kids played with a ball and a bat; Black kids had a stick and a bundle of rags. Most land in the Bootheel was not available to Jim Sr. and the other Black farmers, only ground in the floodway seemed to be for sale. And even there, the higher, drier acres were off limits. Jim Sr. and his partners could buy only cut-over cypress swamp, just stumps and mud. Looking for a tax break, the big logging companies were selling it cheap. While the land was initially advertised at $16 an acre, the men were eventually hustled into paying $28.

Undeterred, Jim Sr. and five other Tennesseans bought eighty acres, and Pinhook began to take root. The logging companies might have made off with the easement checks, unless there never were any; following a 1930s lawsuit, the Corps didn't buy easements on about twenty thousand acres that were vulnerable to backwater flooding. The Tennesseans couldn't afford to care. They seized the chance to own land.

As she drove her client home from the dentist, Twan tried to calm down. It was only backwater, she reassured herself. Backwater has no current. It fills in slowly, like a lake. It has no destructive force.

The eight Robinson kids had grown up with backwater. A gap had been left in the levee at the floodway's downstream end. Jadwin's plan called for closing it, but environmentalists objected: Closing the gap, they said, would drain hundreds of acres of wetland. Alongside his white neighbors, Jim Jr. fought without success to get the gap closed. Year after year, the Robinsons watched water creep over their lawn and inch toward their house. Once in a while, it made it up the steps and into the first floor. Pinhook, located on slightly higher ground than the surrounding fields, would become an island. People rode tractors through the water to catch the school bus or drive into the nearby town of East Prairie to shop. Twan imagined that the 2011 flood would be no different. They would stay with family in nearby Sikeston, or come and go by tractor the way they used to.

Debra also called to warn her sister LaToya and her mother, Aretha, that the levee was going to be blown. At the time, mother and daughter were selling plates of chicken in Cape Girardeau, at AT&T, where LaToya worked, to raise money for a new roof on their church. The brick Union Baptist Church and the community center behind it were the only public buildings left in Pinhook, which had declined over the years. It no longer had a store or a school. Twan, Debra, and Aretha each had a house within sight of the church.

After college at Southeast Missouri State University, Twan had moved to Michigan, where she worked in student housing on the Michigan State University campus in East Lansing. She met people from all over the world and liked the way they treated one another. When that job ended, she came home. She loved living close to her mom and sister, but she was ambivalent about being back in Missouri. One day in East Prairie she heard someone talking about "colored people." She had heard this kind of talk all her life, but this time was different. She resolved to leave Mississippi County, but—for now—a home in Pinhook was all that she could afford. By 2011 she was thirteen years into a thirty-year mortgage.

Growing up, Twan and her siblings knew it wasn't safe for them to be in East Prairie—the closest town to Pinhook—after dark. Nevertheless, Jim Jr. and his father became brokers of the county's Black vote. Democratic candidates called at their house in Pinhook, promising favors in return for support. The Robinsons backed candidates who they hoped would treat people equally.

Around Pinhook, the Black landowners cleared the stumps left by the loggers, dug ditches to drain the ground, graded and broke the soil, and turned the former cypress swamp into productive and valuable land. At one time, Jim Jr. owned or rented thousands of acres, but after he died much of the land was sold off. Few Pinhook residents farmed anymore. Jim Jr.'s generation was getting old, and it seemed like Twan's generation wasn't interested. Her brothers had moved away to work in other places; one was in New York.

Twan dropped off her client in New Madrid and drove home. She turned off Missouri 80 and onto Route AA. As she got deeper into the floodway, she noticed that water had covered the fields south of Pinhook, but that seemed normal enough to her for a rainy spring and a high river.

She met Debra at their mother's house. Aretha was seventy-one and needed a walker to move around. She had lived in Pinhook since she was sixteen. The house Jim Jr. built for her and their eight kids stood on four acres. It had a stone façade, two gables, five bedrooms, and a gazebo. She had filled it with a lifetime of possessions, and now her daughters had to pack up a little of that lifetime before the flood washed it away.

"She wanted it all," Twan said.

They loaded as many of Aretha's things as they could into a semi-trailer on loan from a cousin's employer. Only then did Twan go to her own house. Her brothers had driven to Poplar Bluff, eighty miles away, to rent U-Hauls. It was dark when they returned. Twan collected a few pieces of furniture and the clothes she wore to work. After the safety of her family, her priority was keeping her job. That night, she slept at LaToya's place in Sikeston, assuming she'd stay a week at most. No one knew when the levee would be blown. And the backwater was rising.

Twan briefly returned to Pinhook the next morning. She went to her file cabinet and stuffed her Social Security card and other important documents into a briefcase. In her agitation and distraction, after packing various things into her car, she left the briefcase on the table. "You are thinking but you're not," she explained later, "just remote controlling." By the time she remembered the briefcase, she had driven out of the floodway and past a National Guard checkpoint on the levee. No one was being allowed back in.

Days earlier, Twan had noticed farm equipment—combines, tractors, planters—driving west, heading out of the floodway. It was planting season. Why were farmers moving equipment out? In Pinhook, nobody knew. Maybe it was the backwater, they thought, nothing important. They didn't ask, and no one stopped to tell them. "We were always treated differently because we were Black," said Twan. In retrospect, she and others wondered if the white farmers had information from the Corps or other government agencies that the people of Pinhook weren't privy to. How else did the white farmers know that the floodway was about to be evacuated?

Debra found out about the evacuation on Facebook. She called local emergency management only to be told that everything was fine, to "stand down." Nothing was going to happen, there was no need to pack up and move. Then her brother called from New York. "You get down there, get your shit, and get out," he said, "because I'm hearing it here in New York City that you need to get out—it's a mandatory evacuation." Debra may have heard the news from coworkers, too, or on the radio at work—she's not totally certain—but she is sure that no one from the Army Corps, the county, the state, or FEMA ever called, or came to anyone's house in Pinhook. Before the 1937 activation, the Corps dropped leaflets from an airplane, telling people to get out. Nothing fell from the sky now but rain.

Seven years later, Twan and Debra remained bitterly certain that they were excluded from communications between a largely white government and the floodway's white farmers.

"It felt like we didn't matter," said Twan.

"Because you were Black?"

"Why else?" she said.

Blowing the levees in 1937 successfully mitigated the destructive power of the flood. According to the Corps, it reduced the flood crest at Cairo by 3.5 feet. However, the human and emotional impact was harder to measure. Testifying before the House Committee on Flood Control in the aftermath of the flood, chief of engineers Major General Edward Markham said, "I am now of the opinion that no plan is satisfactory which is based upon deliberately turning floodwaters upon the homes and property of people, even though the right to do so may have been paid for in advance."

Congress appeared to heed Markham's words in 1938, when it authorized construction of a massive dam across the Tennessee River, a tributary of the Ohio. Tributary dams were recognized as another way to manage floods in a river basin, and the government built dozens of them. The more water held upstream, the lower the crests downstream. The dam built on the Tennessee River created the largest reservoir east of the Mississippi. The nearby Cumberland River was dammed, too, creating Lake Barkley, and a canal was dug, effectively combining the two lakes into one massive reservoir, holding back the Ohio's two biggest tributaries. The dams were owned and operated by the Tennessee Valley Authority, but their purpose was to protect the Ohio-Mississippi confluence. Though it was never stated outright, the Corps hoped that the dams would drastically reduce the need to blow the Birds Point–New Madrid Floodway.

As the Ohio Valley snowpack melted in the late winter of 2011, widespread rains drenched the region. Unseasonably warm temperatures melted the remaining Ohio Valley snow in forty-eight hours. A flood rose on the Ohio. Snow began to melt across the Upper Midwest, where the ground, already saturated from the year before, could not absorb the runoff. All that water entered the Upper Mississippi while rain continued to fall across the Ohio basin. The two bulges of water were forecast to reach the Mississippi-Ohio confluence simultaneously. Then a frontal system stalled over the Ohio and Arkansas Valleys. And the rain didn't let up.

On April 21, in a forecast that was not made public, the National Weather Service informed the Corps during a call that the Ohio could hit 61.1 feet at Cairo on May 3, or early May 4. The river had never been that high, and the Corps officials knew it. The upper fuseplug of the Birds Point–New Madrid Floodway, built to a height corresponding to 60.5 feet on the Cairo gauge, had been fitted with subterranean pipes that could be pumped full of liquid explosives. Everyone on the call knew that 61 feet at Cairo was the "trigger point" for blowing the levee and inundating the floodway.

The forecast held, and four days later the plan was put into motion. In Memphis harbor, work crews began loading two pump barges, each with 192 barrels of aluminum powder, six 2,500-gallon tanks of liquid blasting agent, two mix-pump units, and two forklifts. An equipment barge was loaded with two bulldozers and two backhoes.

That same day, the colonel of the Corps' Louisville District began making release decisions for the Tennessee Valley Authority's reservoirs, including Barkley and Kentucky Lakes. The Corps was authorized to take over management of the TVA dams when the Ohio rose above forty-five feet on the Cairo gauge. The colonel wanted to get as much water out of the reservoirs as he could to create room for the coming deluge. By dumping the reservoirs, he was raising the river, but he was doing so before the crest arrived. Ideally, the TVA reservoirs would have enough space to hold water at the crucial moment, lowering the crest just as a floodway would.

On April 25, Easter Monday, the river stood at 56.5 feet at Cairo, up two feet from the day before and higher than the Great Flood of 1927. That night, Debra was watching TV with her mother and sisters when a newscast mentioned the rising river and the possibility of blowing the levee. She saw the Facebook post the next day, talked to her brother, then began warning everybody in Pinhook that it was time to evacuate. Backwater was already on the roads. Soon, their cars wouldn't be able to drive out.

During the next four days, sand boils the size of houses appeared in downtown Cairo, inside the city's thirteen-foot floodwall. Dozens of Corps personnel, thousands of sandbags, and a lot of heavy equipment were mobilized. The city of almost three thousand declared a voluntary evacuation, then a mandatory one. Boils attacked the levees downriver in Fulton County, Kentucky, and Dyer County, Tennessee. In Caruthersville, Missouri, population six thousand, the Mississippi rose to within a foot of the downtown floodwall's top.

Above all, the Corps had to protect the Commerce to Birds Point Levee, shielding 2.5 million acres from flooding, including the towns of Charleston, Sikeston, New Madrid, and East Prairie. A single breach here would cause destruction far eclipsing every possible loss in the floodway. Besides, the landowners in the floodway had already been paid.

The Mississippi River Commission is responsible for the entire MR&T Project, including the floodway. The decision to blow the levee—or not—rested in the hands of the commission president, Major General Michael Walsh. On April 27, Walsh and his staff attended a town hall meeting in East Prairie with Republican Representative Jo Ann Emerson. "There's a lot of concern and rumors, and some of the rumors are creating panic," Emerson told the crowd. The floodway's two hundred or so residents were now under orders to evacuate. The National Guard had set up checkpoints on the levee, and the sheriff's office would be going door-to-door, making sure people were out. Still, nobody knew exactly which areas would be flooded or when. The National Guard had mistakenly tried to evacuate Anniston, a small town outside the floodway. Anniston was safe, the crowd was assured.

Guarded and unflappable, Walsh stood there in his combat fatigues, just inches from seventy-five angry locals. Meanwhile, Emerson produced a photocopied page from *Upon Their Shoulders*, a history of the Mississippi River Commission, and read aloud General Markham's quote about never turning floodwaters upon homes or property. The crowd cheered. "We're going to do everything possible to prevent the Corps from operating the floodway," Emerson said, to more cheering. Then she turned the floor over to Walsh.

One of the authors of *Upon Their Shoulders* was in the audience, too. Charles Camillo, then the commission's official historian and assistant executive director, served as its liaison to the civilian world, part fixer, part concierge. Camillo was with Walsh and the other officials every day that fateful spring and took careful notes, which became the book *Divine Providence: The 2011 Flood in the Mississippi River and Tributaries Project*. In that hot room with those angry farmers, Camillo felt sorry for his boss. He could see that Emerson and the farmers were trying to make Walsh's decision a hard one—to make it personal, to put a face on it. Camillo wrote that Emerson's plan to make Walsh uncomfortable "worked initially. His face reddened. His voice wavered when he started to speak. . . ."

But Walsh soon recovered his composure and began detailing the Corps' efforts to hold floodwater in various tributary basins. The river at Cairo was expected to remain at 60.5 feet or higher for the next eight or ten days. "That equates to about 2.5 million cubic feet per second of water coming down the river; that's a huge amount of pressure on a system that has not seen this type of pressure since we built it," Walsh told the crowd. "We're all fighting for inches."

"It seems to me the Corps is going to give us up for a few inches," remarked one East Prairie resident.

Kevin Mainord, the mayor of East Prairie, stood up to say that when he was in high school, in 1973, locals used sandbags and muscle to fortify the levees and fend off rising water. Why couldn't they flood-fight now at the fuseplugs?

Walsh said that sandbagging would do nothing to help at this point.

"You're not giving us a chance to save our livelihoods," said Mainord, who farmed in the floodway and helped manage a family-owned seed and fertilizer dealership.

The crowd wasn't just angry that their homes and land might be flooded—they were also angry because they didn't think they were being told the truth, Mainord said later. Even Mainord, an elected official, didn't know what the Corps was planning. Like everyone else, he learned about it from the press. To him, Walsh seemed unwilling to divulge anything more than he had to, a military man focused on the mission—in this case, to blow up a levee. Mainord echoed a sentiment heard throughout the region: This was an experiment, and Mainord and his neighbors were white mice. Did Walsh and his staff truly consider the people whose lives and livelihoods would be affected? "I didn't get the express feeling that they cared," said Mainord.

No one from Pinhook attended the meeting. They never heard about it, even though they had the strongest argument against blowing the levee. They never received any flowage easements, and hadn't had an authentic choice about settling in the floodway to begin with; it was the only ground they were allowed to own.

Most people seem to have learned about the evacuation from the media or from Facebook. Though by April 25 the Corps had established a joint information center in Sikeston where it held daily and sometimes twice-daily press conferences, what no one in authority seemed to have known, or to have accounted for, was when the roads to Pinhook would go under water. If Debra had waited for an official notice to evacuate, anyone sent to her door would have had to arrive by boat.

John Story, a white farmer, had an office and eleven grain bins five miles east of Pinhook. His family had farmed in the area for generations and owned thousands of acres in the floodway. Story learned that the levee might be blown from TV news, but as president of the local drainage district, he well understood the probability of drastic action. Decades of levee-related activism had taught him what to watch for. He knew that the Corps would start loading explosives in Memphis when the river rose into the fifties on the Cairo gauge. As soon as it did, he decided to move his rolling equipment—worth millions—to a farm he owned beyond the floodway. He loaded up two 53-foot trailer trucks with his files, his computers, furniture from his office, tools from his shop, and the belongings of five farmhand families who lived in houses he owned in the floodway.

It might have been Story's equipment that Twan Robinson saw driving past Pinhook that day, when she wondered if the white farmers knew something that she didn't. Her father, Jim Jr., had understood the river as well as Story. If he had been alive in 2011, Pinhook might have suspected the Corps' intentions sooner.

Story wasn't deterred by the checkpoints on the levee. Well-connected people like him went right on moving their belongings out. Jim Jr. would likely have reacted in the same way. "People were like, 'Screw you, I've been on this place my entire life and you tell me I'm not going to my house?'" Story said. "People were starting to get their own guns. It was getting ugly."

As the levee neared its limits and floodway activation loomed, the media focused on the acreage that white farmers stood to lose. Pinhook was barely mentioned. The line repeated in news reports was 130,000 acres "and a few homes," Twan remembered. "We were 'a few homes.'"

County sheriffs ordered a mandatory evacuation of the floodway on April 27, and seven hundred Missouri National Guard members were called up to help. As farmers resisting the evacuation made off with their last truckloads, the sheriffs did a final sweep. Paul Haney and his wife were at home when a deputy knocked on their door. Haney, a seventy-two-year-old white man, had grown up in the floodway and lived there his entire life, except the years he was in the military and deployed to Vietnam. His house was on a rise just north of Pinhook; an old friend of the Robinsons, he was particularly close to Jim Jr.'s brother Lynell. Haney's father had endured the activation of 1937—a lifetime ago. He had heard rumors that the Corps might blow the levee in 2011, but those rumors would fly whenever the river rose, said Haney; he ignored them.

The deputy at the door told Haney that he had to leave by 4 p.m. That gave Haney and his wife eight hours to pack up everything they owned. Haney was livid. It was humiliating, offensive, to get run out of your home, where you'd lived all your life, by the Sheriff's Department. They had managed only to load the backseat of a car when the deputy returned. This time, he intended to escort the couple out. Haney said he wasn't finished packing. The deputy put his hand on his pistol and held it there. Haney got the message, and he and his wife left peaceably.

On April 28, an eleventh-hour lawsuit challenging the Corps' right to operate the floodway was settled in the Corps' favor. The State of Missouri claimed that, by knocking over and spilling all the fuel and chemicals left in the floodway, the Corps' actions would violate the Clean Water Act. Judge Stephen N. Limbaugh Jr. (Rush Limbaugh's cousin) of the U.S. District Court in Cape Girardeau ruled that while the pollutants might spill, the threat to the nation's water was not sufficient to stop the Corps.

By April 30, the evacuation was complete. Across the basin, almost every reservoir had reached capacity. Kentucky and Barkley Lakes were full and had begun to release water. The Corps was out of options. Within the command structure, Walsh alone still balked at blowing the levee.

He wasn't yet sure it was necessary. The State of Missouri had also asked the Supreme Court for an injunction barring the Corps from activating the floodway. For a day or two, Walsh hoped that the high court would decide for him. It did not. Missouri's request was denied. Walsh's superiors briefed the Office of the President: Would the executive branch like to decide? There were Democrats on one side of the river and Republicans on the other; the White House demurred.

On May 1, the sheriffs checked the floodway a final time, then flew over it with thermal imaging equipment to make sure no one remained. That morning, the district colonels warned Walsh that area levees could not hold back this flood. They begged him to go to H minus 3: separating the barges and pumping the explosives into the fuseplug. Cairo stood at 59.9 feet. Rain fell by the inch. That afternoon, after seeking additional input from weather experts and colonels, Walsh gave the order: H minus 3.

As he went to bed that night aboard the Motor Vessel *Mississippi*, Walsh was fairly certain that they wouldn't have to do it. There had been a break in the rain, and no more was forecast. He thought, "We got this beat." Three hours later he woke to thunder. Rain lashed his stateroom windows. Walsh had visited the levees in Cairo and in Fulton County; he'd seen the boils, and he was now confident that if he didn't activate the floodway, some levee somewhere would fail.

Walsh didn't know it, but a levee was failing that night. Above Cairo on the Mississippi, across from Thompson Bend, thousands of acres and several small communities were deluged when the private Len Small levee gave way. In the nearby town of Olive Branch, Illinois, population eight hundred, a third of the homes had water in them when Walsh woke up. If he had acted sooner, Olive Branch might have been spared.

In the morning, Walsh started calling the governors. Lightning forced a twelve-hour delay, but on the afternoon of May 2, the blast site was ready. Charles Camillo was in the room when a colonel addressed Walsh: "Sir, I am requesting permission to blow the levee. The ERDC team leader is ready to go hot."

"Approved" was all Walsh said.

At ten that night, several Mississippi River Commissioners were on the setback levee, facing the upper fuseplug. Everything was in position. "Fire in the hole! Fire in the hole! Fire in the hole!" Walsh bellowed. "Five. Four. Three. Two. One."

Twan Robinson waited with her mother and sister LaToya at her brother's house in Sikeston. Twan had made a sign that said "Homeless" in marker on white paper. The detonation aired live on TV. In quick succession, orange explosions lit up the night sky, backlighting the long, flat levee. Almost thirty miles away, the Robinsons heard the boom, then felt the aftershock. Knowing that the floodwaters were raging toward their homes, Twan held up her sign and laughed. Now it was true.

As the levee disintegrated in a cloud of earth, four hundred thousand cubic feet of water per second—the volume of four Niagara Falls or twenty Hudson Rivers—took a right-hand turn and swept violently into the floodway. The torrent scoured away earth and deposited sand. The water was four feet high in Paul Haney's house, and up to the light switches in John Story's office. In Pinhook, it swirled around the eaves of the single-story homes, burst through windows and doors, ripped off Sheetrock by the yard, and filled rooms with sand and mud a foot deep. After the water receded, Twan found a napkin stuck flat to her ceiling.

Animals drowned, or ran to high ground. The fate of wildlife seemed to disturb Haney more than the defilement of his land or home. There were dead rabbits, bobcats, mice. Hundreds of songbirds that would not fly away ultimately starved to death, he said. "You can imagine them poor old deer standing out there in the water, holding their noses up, and the water still rising, until they all drowned."

In total, 133,000 acres were inundated, about a hundred buildings were destroyed, and more than two hundred people, Black and white, were displaced. Most never returned.

The Corps had breached Lester Goodin's levee. The fuseplug was his levee district's responsibility, though for three decades he had fought against its use. Before him, his father and uncle had fought—locally and in Washington—to get the floodway decommissioned. They lobbied, testified, made speeches, filed lawsuits. It was "this sword of Damocles hanging over our head all these years," Goodin said. "We were trying to get rid of the sword, but I never thought the thread would break. None of us ever thought we would see it."

Goodin attended one last meeting with the Corps on the night the fuseplug was detonated. He drove away in the rain, knowing he was beaten. That night was miserable, almost wintry, and the weather matched Goodin's mood. He didn't want to witness the explosions, and drove home to Cape Girardeau. He was awake at 10 p.m. but didn't feel the blast, though many of his neighbors said they did.

Before the levee was blown, the Weather Service had predicted a 63.5-foot crest at Cairo on May 5. Instead, the Ohio fell from 61.7, at the moment of the breach, to 59.6 on May 5, almost four feet lower than forecast. The Corps credited the river's drop to the activation of the floodway. Just about no one in the Bootheel believed this. Goodin was an exception. "I couldn't, in good conscience, say that it is absolutely worthless, although I really thought it was," he said, drinking coffee in his book-filled living room. Inundating the floodway, he admitted, did lower the crest "pretty much according to prediction." Of the people who question that, who say the Ohio would not have risen any higher, Goodin said: "They're wrong. It did lower at Cairo."

The farmers cleared the mud, sand, dead fish, and debris off the roads, and as soon as the ground inside the floodway was dry enough to support a tractor, they ventured out into the fields with their equipment, moving earth and planting around the scour holes. The smell of rotten fish and dead animals lingered for weeks, said Story, who was able to plant 84 percent of his ground with soybeans, the only crop that could mature in what remained of the growing season. Most of the land in the floodway was successfully planted that year, and the crop was better than anyone expected. Farmers got their cash flowing again and began to recover from their ordeal, at least economically. The damage to the land, equipment, and buildings would take years to repair.

Goodin himself lost almost nothing in the flood: "I came through smelling like a rose compared to my neighbors." His land wasn't in the path of the scour. The acreage owned by his levee district, immediately in front of the upper fuseplug, was devastated but restored at taxpayer expense. The activation was less painful for him, and he was able to let it go. "I was angry," he said, "but anger doesn't get anything done."

Seven years later, many of his fellow farmers were as angry as ever. In their eyes, Goodin's conclusion that the floodway worked as intended only served to discredit him.

Paul Haney admitted that, sure, Goodin farms in the floodway, but he doesn't live in it: "He don't have the feel of the land in his blood."

"Up in them damn hills, he didn't care," said Ellot Raffety, a board member of Levee District No. 3. "You got to watch Lester. He's a little off the wall. He's got land there, but he wasn't staying over there." (Goodin and Raffety were well acquainted, having been on the levee board together for twenty-five years. Goodin said that "being around Ellot is like swimming with an anchor.")

A true flatlander, Raffety lived on the floodplain, just across the Mississippi River from Cairo, not up in the hills. He could look out his kitchen window into Kentucky, and out his bedroom window into Illinois. Raffety believed that the Ohio couldn't have gotten any higher at Cairo because water was already overtopping the fuseplug before the Corps blew it up. If not for the detonation, water would have dribbled over the levee into the floodway, the crest would have passed, and everyone would have been fine. "Wasn't anything behind the crest but water in the Tennessee Valley," he said.

Raffety sought to defend his reasoning seven years later, following a public meeting conducted by the Mississippi River Commission. A hollow-cheeked man in pressed khakis, Raffety was up in the pilothouse of the commission's towboat with his friend Charles Davis, waiting for lunch to be called. The Motor Vessel *Mississippi* had left New Madrid and was chugging lazily downstream. Davis, chief engineer of the St. Francis Levee District, based in nearby Caruthersville, Missouri, had just finished testifying about the panic that gripped his town as the water rose in 2011. The river had begun to seep under Caruthersville's floodwall and almost came over the top. People didn't think the wall would hold, and rushed to pack up their belongings. The banks closed. The town spent five days struggling to build a secondary floodwall with tons of earth and gravel, and sixty thousand sandbags. The original government plan called for the floodwall to be at least three feet higher than what was ultimately built. In his speech, Davis asked the commission to keep its promise and raise it now.

If the floodway hadn't been inundated, Davis believed, Caruthersville would have been in grave danger. Raffety didn't buy it. Since water in the floodway reentered the river at New Madrid, he told Davis, it didn't benefit anybody downstream, in places like Caruthersville.

Davis disagreed. When the levee blew at Birds Point, he said, it dropped Caruthersville "1.6, 1.7 feet on the wall."

"Really? That surprises me," Raffety said. He'd assumed that when the floodway's water returned to the river, it would raise the Mississippi downstream.

"It didn't on us," said Davis.

Raffety's mind remained unchanged.

By the spring of 2018, most of the scour holes in the floodway had been filled and the levees had been restored. The landscape appeared verdant and fruitful, as if the torrent had never happened. But the feeling of the place had changed. Houses and stores had disappeared. Cars were rare. The floodway's population of more than two hundred had been reduced to a handful, in three or four homes.

On a gray April day seven years after the flood, no one drove down Missouri 77 and no buildings broke the monotony of trees and fields. Finally, a cluster of grain bins hove into view—first a set belonging to Pete Story, then, after an interlude of farmland, a set belonging to Pete's cousin, John.

John Story's office was in a clean, gray, vinyl-sided building next to a long shop and equipment shed in a town called Wolf Island. The only thing left of the town now was Story's farm. On a flat screen in the office, Fox News was playing on mute. A shaggy dog rested in its bed beside a desk stacked neatly with papers, bills, and pay stubs.

Story didn't lose his home—he lived twenty miles away, in Charleston—but many commercial buildings and six houses that he owned in the floodway were destroyed. His farmhands had lived in the houses as part of their pay. Because these weren't primary residences, they were ineligible for FEMA aid. Far from helping him rebuild, it seemed to Story as if FEMA were trying to stop him. If a structure's flood damage was greater than 50 percent of its value, FEMA would not let the owner rebuild unless the lowest floor of a new building was above the one-hundred-year floodplain, an imaginary line in the air that supposedly represented the crest of a flood that had a 1-in-100 chance of occurring in any given year. The government essentially told Story that he would have to put everything up on stilts. He walked out of a meeting with FEMA's representatives, telling them, "I'll be in Wolf Island putting my shit back together. If you have a problem, come see me." FEMA never came. Since the flood, Story calculated, he'd spent $1 million rebuilding. Without flood insurance. Without loans.

He did collect crop insurance for some corn and winter wheat that washed away in 2011, though even that money was initially uncertain. His policies covered only crops damaged by acts of God, the insurance companies told him, and blowing the levee was an act of man.

Story got up from his desk, walked into another room, and unlocked a safe. He withdrew a thick binder full of documents, then flipped through the folders to find a typewritten letter. Addressed to his maternal grandparents from the Army Corps of Engineers, it read:

Dear Mr. and Mrs. Dernoncourt:
    Submitted in payment for the purchase of a Flowage Easement over Tract No. 429 E, Birds Point-New Madrid Floodway, Mississippi River and Tributaries, is U.S. Treasurer's Check No. 178, 179, dated 22 January 1975, in the amount of $50.00. A copy of Payment and Closing Sheet and Receipt for United States Treasurer's Check covering this transaction is inclosed [sic] for your files.
    Your assistance and cooperation in this matter are greatly appreciated,

"In 1975, fifty bucks?" scoffed Story, who now owns the 320-acre Dernoncourt farm. "Come on!" For $50 the government had bought the right to destroy this land—and for what?

Almost every white farmer in the region repeated two arguments against using the floodway: first, that the crest had passed and blowing the levee was unnecessary; and second, that the farmland in the floodway was worth far more than the city of Cairo, north of the floodway and across the river in Illinois. In 2011, Cairo was 70 percent Black, and 50 percent of its population lived below the poverty line. "Have you seen Cairo?" Story asked. The implication: If you'd seen Cairo, you'd know it wasn't worth saving.

Cairo was a substantial and prosperous town before, during, and after the Civil War. Its position at the Ohio-Mississippi confluence made it important both strategically and commercially. But the steamboats disappeared, then the railroads stopped stopping; corruption, mismanagement, and racial and civil unrest followed. Many of those who could afford to leave left, and the city's population declined by almost 90 percent. In downtown Cairo, two streets of handsome brick storefronts stood abandoned for years before most were demolished. Illinoisans joke that money in their famously mismanaged state trickles down from north to south. Chicago soaks up a lot, and very little seeps below Springfield. Travelers crossing into Southern Illinois from other states are usually greeted by truck stops and strip clubs. Cairo doesn't have either, though it does have some grand buildings testifying to its former affluence—the turreted brick library, the Italianate Old Customs House, and the fourteen-room Magnolia Manor.

Meanwhile, across the Mississippi in the floodway, a vast swamp with a few rough farms had become an agricultural machine, cranking out millions of dollars in soybeans and corn every year—the most productive farmland in Missouri, Story said. "Not that I have anything against those people at all," he continued. "The town just didn't work. The town that we built this system to protect no longer merits protection, in my mind, if you're trying to say that we're going to spend all this money to protect the greater good, who is the greater good now?"

The 1928 Flood Control Act does privilege Cairo, which still had nearly three thousand residents in 2011. It's hard to imagine any metric that would value farmland, no matter how productive, more than a historic city of this size, even with its homes and businesses evacuated. The white floodway farmers like to pit themselves against the mostly Black city across the river, but officials from a dozen other communities in four states acknowledge that Walsh's decision to operate the floodway helped their towns survive the 2011 flood. An unplanned levee breach almost anywhere would certainly have been worse, in human and economic terms, than the controlled operation of the floodway.

"The Corps will tell you all kinds of lies," Story said. He took down a signed copy of the official history of the 2011 flood, Charles Camillo's *Divine Providence*. Story opened it to a graph showing the height that the Ohio would have reached at Cairo if the floodway hadn't been breached: 63.5 feet. Camillo had given the book to Story over dinner one night on the Motor Vessel *Mississippi*. Story didn't believe the graph. He had asked the assembled engineers how the river could possibly rise to 63 feet when the fuseplug levee—already overtopping at low spots before the floodway was operated—would have overtopped at 60.5 without being blown. Wouldn't the river have remained at 60.5? If you fill a bathtub, he reasoned, the water can rise only as high as the side of the tub. How could it get higher?

Scientists from the University of California at Irvine, Ohio State, and the University of Bristol concluded in a 2015 paper that, if the fuseplugs had remained intact, the river would have risen 2.6 feet higher before spilling over into the floodway. The detonation, they wrote, "greatly reduced risk of levee failures" around the confluence area, though it caused about $50 million in additional damages.

The dining engineers gave Story an explanation, but he wasn't satisfied.

Didn't blowing the levee spare Caruthersville?

"Bullshit," Story said. "All the Corps guys drink the Kool-Aid and say what they got to say." It's about money and self-preservation, in his opinion: The Corps has to spend every penny it gets from Congress every year, so its budget won't be cut. "I don't see how blowing this levee did anything at all," Story said. "And I never will."

A few miles away, Paul Haney's one-story brick home was an island of meticulous housekeeping in a sea of farmland and trees. The front door was unlocked. Inside, Haney sat chain-smoking in a soft chair with the TV on and muted. He had retired from farming two years before. Now almost eighty, he was still an imposing presence, more than six feet tall. His wife had died not long after the flood. She had decorated their living room with family photographs and paintings of cherubic children.

"Have you been to Cairo?" he began. It was Story's argument all over again. "Do you see that it's worth blowing the levee for thousands of acres of land and all the wildlife?"

Haney had lost his property abstract in the flood, but he didn't recall any mention in it of a flowage easement. When he was allowed to return to his sodden house, he knew that he would have to tear out the drywall and carpets, and take the walls down to the studs. He had no flood insurance, and couldn't qualify for a loan. "I had the money to build a house, and that's what I done," he said. The same house in the same spot. Perfectly restored.

Haney said he did get $29,000 from FEMA, the amount offered to "everybody that could prove ownership of their house. See, if you had any kind of loan on it then you didn't own it." If his neighbors in Pinhook had mortgages, he said, "the government, instead of paying them, paid the bank." The road to Haney's house was breaking off in chunks. The highway department no longer maintained it. Past his home, down the slope, the broken blacktop led straight to Pinhook.

There, FEMA found, every house except Lynell Robinson's had damages exceeding 50 percent of its value. To legally rebuild, homeowners would have to elevate their houses on twelve-foot mounds or stilts. The only affordable way to rebuild in the floodway was to pay cash and defy FEMA, as Story and Haney had. No one from Pinhook could afford to do that, and besides, they needed FEMA's help. The village would survive in spirit, but the place would have to be abandoned. Several residents had flood insurance, but if they had mortgages, their banks took the payout. Similarly, it appears that only those without mortgages or insurance received direct aid from FEMA. Twan Robinson's flood insurance payment went to the U.S. Department of Agriculture, which held the mortgage on her house. Uncle Sam had to get paid first, she would say, "even though Uncle Sam tore up all of that."

Once the flood subsided and the National Guard left, Anthony Schick of the *Columbia Missourian* reported, "Thieves entered Pinhook and took anything of value, anything metal. They took the garage door off Aretha's house, as well as the windows and copper wiring. They took furnaces, refrigerators and sinks. They took the bell from Union Baptist Church." The church and houses stood abandoned, curtains flapping in the wind, walls crumbling, until the community paid to knock them down—all except Lynell's. "He said he wouldn't never leave," Twan said.

Sociologist Junia Howell, a professor at the University of Pittsburgh, studies the intersection of income inequality and natural disasters. She has identified "three dimensions of social stratification": education, home ownership, and race. When natural disasters strike, these three factors determine whether an individual prospers or falters, profits or loses everything.

In *Damages Done: The Longitudinal Impacts of Natural Hazards on Wealth Inequality in the United States*, coauthored with James R. Elliott, Howell wrote that "whites who lived in counties with very little hazard damage ($100,000) over the 1999–2013 period gained, on average, $26,000 in wealth. By comparison, similar whites living in counties that experienced $10 billion in hazard damage gained nearly five times that much, or $126,000. For Blacks, results cut the other way. Those who lived in counties with just $100,000 in hazard damage gained an estimated $19,000 on average; whereas those living in counties with $10 billion in hazard damage *lost* an estimated $27,000."

Similarly, Howell's research revealed that "the more FEMA aid a county receives, the more unequal wealth becomes between more and less advantaged residents. . . . whites living in counties that received $900 million in FEMA aid during 1999–2013 accumulated $55,000 *more* wealth than otherwise similar whites living in counties that received only $1,000 in FEMA aid. Conversely, Blacks living in counties that received $900 million in FEMA aid accumulated $82,000 *less* wealth than otherwise similar Blacks living in counties that received only $1,000 in FEMA aid."

One reason that government aid is not equally distributed is that FEMA's disaster relief program focuses on property, a form of wealth that many people don't have. Relief for renters is much more limited. "Natural hazards do not just bring damages, they also bring resources," wrote Howell, and "equal aid is not equitable aid, especially when it is systemically designed to restore property rather than communities." In this way, FEMA rewards wealthy more than poor, white more than Black.

Imagine a disaster-relief system that quantified the value of a community and then paid to restore it, instead of focusing on single pieces of private property. Again and again in interviews, Debra Robinson-Tarver said, "We're not out to get more than what we had. We're just out to get what we had back." Howell's work shows that white people like Story and Haney benefit from aid and buyouts, while Black people like Debra and her sister Twan become poorer overall.

From 1980 to mid-2020, the United States experienced 273 weather events—everything from storms to wildfires to floods—each of which resulted in more than $1 billion in damages. Severe storms, tropical cyclones, and flooding cost the most. Between 2004 and 2015 alone, FEMA distributed $146 billion in disaster relief. If whites gain and Blacks lose, how has this massive outlay contributed to the social disaster of inequality? Howell wrote that under the current system "hazard recovery is understood not as a unified act of resilience but as a struggle by privileged residents to restore the local social order, as well as opportunities it presents."

Several Pinhook residents were college-educated and most owned land. According to Howell's three dimensions of social stratification, they had just one disadvantage: they were Black. Whether or not race directly influenced how the Corps handled the 2011 floodway evacuation and activation, or how local, state, and federal emergency management responded, an aid system was already in place that automatically helped the white and the well-to-do.

On a rainy April day seven years after the flood, Twan crossed over the levee into the floodway and drove down the road to Pinhook. A square of brown grass and three budding trees marked the site of her former house. Rings of abandoned shrubbery described the footprints of other homes. Grass had grown over their collapsed foundations, covering all but a little rubble. Half the roads were paved, half not. One intact foundation remained, one shed, one boarded-up house, and her uncle Lynell's home.

The village owned this lonesome land, though no one could legally build on it. It was hard to find people to mow, to keep the weeds back. Those with the time were elderly or disabled. The others worked every day. The Robinson sisters tried to erect a pavilion for the annual Pinhook Day celebration, even securing a grant for the purpose, but before they could sort out a design and untangle the floodplain restrictions, the grant expired. They did put up a sign that read: "Pinhook Reclaimed 2017."

A relative of Twan's had turned the town's streets into a parking lot for farm equipment: rows of long claws on wheels, zigzags of metal, a fifty-three-foot trailer. The villagers had objected—there was something disrespectful about blocking access to a repository of so many memories—but the relative ignored requests to clear the streets.

Aretha Robinson's gazebo, a memorial to Jim Jr., stood in one piece, a little crooked. Before the flood the local VFW had rescued a plaque honoring his army service. The concrete floors of her garage and patio remained below a weathered basketball hoop.

A piece of farm machinery blocked the curved drive that had once led to the Union Baptist Church. Two sections of yellow metal framework lay where the sanctuary had been.

For seven years, a group of Pinhook residents—mostly members of the Robinson family—struggled to reestablish the village, to resist the erasure of their community. They pressed the government to help them rebuild. Debra, the unofficial leader of Pinhook, filled out forms while the streaks of gray in her hair multiplied. Twan's studies left her little time to help. At the end of 2013, she completed a master's in social work. Three years later, she became a licensed clinical social worker and got a job as a behavioral health consultant.

Much of Debra's paperwork was routed through the Bootheel Regional Planning Commission, which applied for funding on Pinhook's behalf. In September of 2011, planning commission officials told the press that FEMA had offered to buy out the village for $1.7 million. A year later, they said that an additional $1.4 million was being sought from the Department of Housing and Urban Development (HUD).

There would be enough to tear down the ruined village and ideally build a new one.

The people of Pinhook began looking for a site that would meet FEMA's specifications. It needed suitable sewer, water, and electricity access, and had to be outside the one-hundred-year flood zone. The Robinson sisters thought that forty acres would be a nice spread for eleven homes, a church, and a community center. They found several promising sites around the Bootheel, but utilities weren't always available. A lot of Mississippi County is farmland, and more than a third of it falls within the one-hundred-year zone. Most acreage outside the zone was too expensive. Some sellers seemed ready to make a deal, but later changed their minds. Twan wondered if her skin color had something to do with it. Location after location fell through or was rejected. Neither the $1.7 million nor the $1.4 million ever materialized, and the Pinhook diaspora began to fray as elders passed away and families gave up on the notion of a new village.

Twan and Debra, both in their fifties, continued living together with their mother in a green one-story frame house owned by a cousin. It was cramped, with little privacy, but it was all they could afford.

On Apache Drive in Sikeston, thirty miles from Pinhook, people began to assemble in the front yard of a nearly finished house. Seven new houses in all were set along the street. The yards had yet to be sodded and the wind picked up the exposed alluvial soil. Members of the crowd shielded their faces against the grit with programs and song sheets. Across the road, farmland stretched to the horizon. The sky spit rain. It was noon on a restless gray day in April of 2018.

The people of Pinhook arrived to enthusiastic hugs, it's-been-so-longs, and thank-Gods. Twenty members of the Amish community stood along the wall of one house, reserved and quiet, the men in straw hats, the women in bonnets. Everyone seemed to know one another, and there was a collective sense of relief, as if something they'd all endured was finally over. In two weeks they would mark the seven-year anniversary of the destruction of Pinhook. Now, the three Robinson sisters, their mother, and three other former Pinhook residents were about to receive the keys to their new homes.

Nine lots had finally been purchased, but it wasn't a new village; seven lots were within shouting distance of one another in this Sikeston subdivision, another was north of Sikeston, one was in Charleston. Ultimately, FEMA bought out twelve homeowners, plus the church, for sums mostly ranging from $28,000 to $2,500. Four hundred and fifty-two thousand dollars had come through from HUD. Many homeowners had taken out small mortgages to cover cost overruns and additional expenses. Catholic Charities of Southern Missouri and Mennonite Disaster Services oversaw the construction project, which attracted about $50,000 worth of donated materials and more than eight thousand hours of volunteer labor. Teams of Amish workers built all nine houses in four months. Working in shifts, they were bused back and forth each week from Canton, Ohio, ten hours away.

A white man from Catholic Charities, wearing a ball cap and a name badge, began with a speech. A young Black preacher said a prayer. The people of Pinhook stood in a group around Aretha, who sat in her wheelchair. The first house was hers, and it was different from the other eight: Some of the reddish-gray stone that Jim Jr. had used to build their home in Pinhook had been salvaged, and now it decorated the lower part of the small one-story frame house. The Amish sang "Amazing Grace." The Catholic Charities man said, "Aretha, welcome home," and gave her the key. Everyone cheered and headed inside with her, admiring appliances and opening closets.

They performed the ritual at each of the seven houses on Apache Drive: the blessings, the "welcome home," the key, everyone going inside and switching lights on and off. Each of the Pinhook men and women received a Lucite plaque from Catholic Charities, a quilt from the Amish, a Bible, and a bouquet of flowers from the Bootheel Economic Development Corporation. The new homeowners appeared overwhelmed, standing shyly, smiling, blinking back the dust.

The sixth house, across the street, was Twan's. It had two bedrooms, one bathroom, and Whirlpool appliances. She shouted her thanks to God. The seventh house was Debra's.

Every Friday during construction, the people of Pinhook had cooked for the workers, who slept at the Charleston Baptist Association campground north of Sikeston. Now everyone drove up to the campground for a final meal. Debra opened with a moment of silence, a form of prayer she'd learned from the Amish. Then came trays of corn, baked beans, pulled pork, sweet potato pie, ice cream, and coffee.

Debra sat across a plastic table from a young man named Marvin Miller, who had managed the house construction. He wore the traditional whiskers and suspenders of the Amish. Miller had a diagram of the floodway, and he asked her, with a gaze of honest curiosity, to show him where the levee had breached. She pointed at the paper with an uncertain finger. Miller drew two parallel lines across the levee with his pen. That was the breach.

"The whole house shook," she told him. "You could hear glass rattling. They didn't do it in the daytime so it could be seen. They did it at night."

Opposites though they might have seemed, the Black Baptists of Pinhook and the white Anabaptists of Ohio shared a deep faith. Bearded Amish men in black hats and collared shirts chatted amiably through dinner with the Robinson brothers in their G.O.A.T. hats and Coogi sweaters.

After dessert, the Robinson sisters stood and sang a gospel song: "He'll wash your sins away, turn a midnight into day / Do you know him, tell me do you know him."

"Him," Twan said, sitting down. "Him being the foundation makes a big difference in our lives." In her faith, she had accepted that Pinhook had been sacrificed for the greater good. "God did this," she said. "I wish that we had not been forgotten for so long, but it was God's timing, not ours, and I respect that."

* * *

Pinhook should never have been built in the floodway; the authorities should have given the residents greater warning and assistance before blowing the levee and washing away their homes; and the victims should have been justly compensated for what they lost—all this is almost certainly true, but it is also almost certainly true that the floodway worked. With climate change adding more rain to the nation's rivers, flood protection systems will need more floodways, not fewer, operated more often, not less.

As painful as it was for the homeowners in Pinhook and the farmers in the Birds Point–New Madrid and—later that same spring—the Morganza Floodway to be collateral damage, the system of levees and floodways managing all the water from Cairo to the Gulf operated as designed in 2011, and continues to do so. The floodwater went where the Corps wanted it to go. In the Mississippi Valley there were a handful of deaths from flooding on the tributaries and thousands of people were evacuated, but only a small fraction of the land went underwater. Jadwin's levees stood. His plan worked. The MR&T Project proposes to manage a massive amount of risk. The water has to go somewhere.

"My opinion is we don't operate the floodways nearly enough," said Ty Wamsley, director of the Coastal and Hydraulics Laboratory at the Corps' Engineering Research and Development Center (ERDC) in Vicksburg. Socially speaking, the more often a floodway is used, the easier it is to use. Because the MR&T floodways aren't used very often, those who live and work within them don't expect to be flooded. In the rare instances when these crucial relief valves are truly needed, people like the Robinsons, Lester Goodin, and Paul Haney want desperately to protect their investments. Through political and legal means, they try to stop the Corps.

A levee designed to be destroyed only exacerbates the outrage and betrayal felt by locals. Similar structures elsewhere employ different, more "passive" designs. When the Sacramento River rises to a certain height, it flows over a concrete weir into the Yolo Bypass, sixty-six thousand acres of sacrificial farmland across from the California capital. No one lives there, and structures are prohibited. Inundation isn't a tragedy or a surprise because, unlike the twice-used Birds Point–New Madrid Floodway, the Yolo Bypass floods about every other year. According to a paper by Brad Walker, published in the *Journal of Earth Science*, "The flooding is slow and gradual, minimizing damages and allowing sediments to settle on the bypass land to enrich soil." (Walker is a colleague of Bob Criss, the Washington University professor who said that flood-waters do not rage.) For those protected by the bypass, the knowledge that it works automatically, every time, without lawsuits or conflict, provides an important sense of security, wrote Walker, while farmers within the bypass accept that their land will often be temporarily flooded but not damaged. These conditions of certainty and predictability do not exist at Birds Point–New Madrid. There, Walker recommended that the fuseplug levees be degraded to the height originally recommended by Jadwin, and paved so they wouldn't erode. Then the floodway would operate passively about every ten years, and the cause of the inundation would seem to be the high river, not the Corps.

Wamsley has three degrees in engineering, but he isn't interested in mega-dams or super-bridges. For Wamsley, engineering is inseparable from science. Understanding how water moves is, for him, the same as understanding how to work with water. Wamsley's job, as leader of one of the most advanced and well-respected water research institutions in the world, is to comprehend and shape the future. To do this, he spends a lot of time figuring out the past, both what the river has done and the past actions of the Corps.

After a public meeting aboard the Motor Vessel *Mississippi*, Wamsley stood in a hallway talking quietly with Rear Admiral Shepard Smith, a Mississippi River Commissioner appointed by President Obama. Smith was the director of the National Oceanic and Atmospheric Administration's Office of Coast Survey. Under fluorescent lights, framed by heavy bulkhead doors, the two men looked like priests meeting along the cloister to discuss an exciting and heretical reading of Scripture. They were questioning the "project design flood," the founding prin-ciple of the entire $16 billion, ninety-one-year-old Mississippi River & Tributaries Project. The two scientists believed that this piece of doctrine had become outdated, maybe even dangerous.

The project design flood is the hypothetical maximum flood that the MR&T levee system was designed to handle. While the concept dates to the system's inception, the current project design flood was developed in 1955. It begins with three different rainstorms occurring in series over different parts of the Mississippi River Basin. The Corps estimated all this rainfall and mapped the water's course down the river. The height and volume of this hypothetical super-flood determines its flowline, an imaginary line the water would trace as it flows downstream. The Corps revised the flowline upward after the 1973 flood—during which the Morganza Floodway was operated for the first time—and considered raising it again after 2011, but ultimately left it the same. The project design flood has never happened. By volume, the flood of 2011 was considered 85 percent of project design.

The MR&T levees have been under construction since 1928. Each year the project is a line item in the Corps' budget. Even so, several stretches of levee are not yet wide enough or tall enough to contain the water anticipated since 1973. Meanwhile, the climate has changed, and so has the technology used to forecast weather and calculate risk.

"We've designed the whole system around being able to pass that one event," said Smith. While that one event is very bad, it's not "necessarily the most likely bad thing that could happen." The project design flood made assumptions that Smith wasn't comfortable with. The MR&T levees "could fail in a number of ways," he said. The flood anticipates a precise ratio of water coming out of each of the Mississippi's tributaries. What if those rivers flooded in "a different combination that could stress the system in different ways?" asked Smith. "We could be unprepared in ways that we don't anticipate, because we've got this one design criteria."

The project design flood is considered "deterministic." It assumes that after the initial inputs of melting snow and rain, everything will happen in a set sequence. During the flood, nine times more water will come out of the Ohio River than the Upper Mississippi. Less than half of the volume of the Upper Mississippi will come from the Missouri, which will deliver one hundred thousand cubic feet per second (cfs) of water into the Mississippi above St. Louis. That's the hypothesis.

In reality, during June of 2019, the Missouri was pumping four hundred thousand cfs into the Upper Mississippi, four times more than the project design flood anticipates, while the Upper Mississippi was flowing through the Thebes Gap at almost nine hundred thousand cfs, more than three times project design. Luckily, the Ohio was delivering far less, so that below Cairo, where the Lower Mississippi begins, the result was a big flood, but not a super-flood—about 60 percent of project design. But what if the Ohio had supplied its full quotient of water on top of the higher flows from the Upper Mississippi? What if the Ohio had delivered more than project design?

Wamsley and Smith didn't know. It was their job to protect people and property from destruction, yet the rivers were behaving in ways the system wasn't designed for. It could still be operated adaptively; the floodways and backwater areas could be used more often, or in a different sequence, said Wamsley. He was confident that the MR&T Project as built still kept people safe, but he was just as confident that the Corps could build something better if given the chance.

The answer, the two men said, was to create a new model to supplement the project design flood. Instead of a single event, a "probabilistic" model would consider many coincident scenarios and their respective likelihoods. Take all possible snowfall patterns and every historic rainstorm, plus more snow and more rain caused by climate change, and also the maximum flow of every tributary; then include variable soil, water, and moisture conditions—"throw all these different potential scenarios at a big supercomputer and let it crunch," said Smith. Then run those floods and their probabilities of occurrence through a model of the existing levee system.

"With the probabilistic analysis, now you can combine these probabilities with the consequences," said Wamsley, "so that you're not putting all this money upstream that's actually transferring hazard downstream, where you may actually have more consequences—and of course, risk is probability times consequences."

"If you take it one step further, and allow us to model the system in failure," Smith said, "failing would take some pressure off the system, so it may be, in fact, that it's impossible to overtop the levees as they're built, because something else would happen first." As with a floodway, a levee breach in one place lowers the crest somewhere else. This was demonstrated in 2019 when levees up and down the Arkansas River blew out. Since water was flowing over Arkansas farm ground and not into the Mississippi, crests there were lower, and the Morganza Floodway didn't have to be opened.

The federal government wants to know how much money to spend in order to benefit the greatest number of people, balancing risk-reduction and cost. "We could be putting our resources in the wrong place," Smith said. The MR&T Project builds the same levee to protect a thousand acres of farm ground that it does to protect a city of fifty thousand people. In fact, the levee in front of the city may breach before the one in front of the farmland. By not breaching, the levee in front of the farmland transfers risk to the levee in front of the city. This is probably happening now, but nobody really knows.

What prevents the Corps from conducting the probabilistic analysis?

"We have the wherewithal," said Wamsley. "We know how to do it—"

"We just haven't done it," said both men in unison. Not surprisingly, it's a matter of money, political will, and political imagination. Congress, or the very top brass at the Corps, would need to ask for such a sweeping re-analysis. And it wouldn't be cheap.

Price is one obstacle. Then there are the entrenched interests that benefit from the current system, said Smith, "including the engine of 'moving dirt' and the jobs at the districts within the government, the contract jobs, and all the levee boards." (Smith was referring to an Army Corps slogan, "Moving Dirt," belabored by the Memphis District commander in an earlier presentation. If the point is to move dirt, rather than to spend money efficiently and intelligently to protect people, then the district looks like a machine designed to perpetuate itself—something the Corps is frequently accused of being.)

"And if you take a risk-based approach, there's obvious winners and losers," said Wamsley. "And some fairly powerful interests are in those areas where you may have the lowest risk." The political fallout of lowering a levee, even if doing so might save lives, would be torrential.

A probabilistic analysis would add risk-based reasoning to the Corps' much-maligned cost-benefit analyses. For instance, it's probably not in the public interest to protect the thousand-acre farm from a highly unlikely flood, but maybe it should be protected from a fifty-year flood. The city of fifty thousand, on the other hand, might merit costlier protection, like a five-hundred-year floodwall. Hard choices would have to be made, said Wamsley, like comparing the "value of farmland versus human lives versus industry."

"There may be some that are comfortably overprotected," Smith said.

The more forward-thinking elements within the Corps weren't publicly pushing to reevaluate the project design flood because they didn't want to undermine the effort to finish the MR&T Project, but both men felt that flood protection systems, like all infrastructure, should adapt to major social and environmental changes. "We should be able to continually evolve it, right?" said Smith, "if we have a new normal climate pattern." Yet most high-level Corps officials seemed to believe that the last few decades of increased flooding and rainfall were nothing new. "We're in a wet cycle right now," said one levee engineer. Another called the last one hundred fifty years of data insufficient to draw any conclusions from. "I don't want to say climate change," said a third. "I hate that term, only morons say that, it's always changing." Not one official in the course of eleven River Commission meetings referred to "climate change" publicly, though everyone agreed that lately there was more water, and most said that more should be done about it.

The levees don't know or care where all that water comes from. Levees either fail or they stand. Determining the cause of changing weather patterns wouldn't ultimately make much of a difference, said Smith, "unless by understanding the cause we can better predict how long and how far the changes we're seeing will go." Smith spoke loudly to be heard over the hubbub that always precedes a public meeting aboard the Motor Vessel *Mississippi*. He was literally surrounded by engineers who believed that the high water upon which they floated wasn't a sign of permanent change. They didn't believe that the Corps needed to redesign anything, because this "wet cycle" would soon end.

Smith disagreed. The science clearly proved that "man-made causes are at least exacerbating, if not causing the current climate change," he said. But he recognized that the people he worked with, up and down the river, were steeped in the divisiveness of contemporary politics. He knew that, for them, the words "climate change" were a trigger. Those two words instantly moved the conversation out of the realms of engineering and science, and into the world of politics. The Corps' recent report, laying out the evidence for more rain all over the Mississippi River Basin, didn't use the words "climate change." Smith understood why. If that term was included, he said, "Some people would stop reading at the cover."

Scientists like Smith and Wamsley needed to convince their more conservative peers that updating the system to better protect people wasn't political. Convincing Congress would be even harder—the leg-islature's 2018 water infrastructure bill didn't mention climate change once in three hundred thirty-two pages.

"If you were to go back and do it from the beginning, you could probably have a much more nimble and efficient system at a much lower cost," Wamsley had said eagerly. "And now we have the computer power to really be able to do that." We know one thing for certain, he said: The project design flood won't happen. If it did, "That'd be a miracle."

The MR&T Project is analogous to the pre–Hurricane Katrina levee system around New Orleans. New Orleans flooded for a lot of reasons, but a probabilistic model and the will to implement it might have saved thousands of lives and billions of dollars. Just like the MR&T Project, the old Hurricane Protection System for Southeast Louisiana was based on a hypothetical event called the standard project hurricane, or SPH. But, unlike the project design flood, it wasn't the worst imaginable. In fact, it was quite likely to occur.

The Corps designed the project hurricane in the 1950s as a bench-mark for determining levels of protection. Yet, over time, perception of the standard shifted from "a general indicator of threat levels to a guarantee of safety," wrote the Independent Levee Investigation Team, a group based at the University of California, Berkeley, that studied the engineering failures that left New Orleans flooded. The methods used to define the project hurricane were forgotten, they wrote, "along with their potential flaws and questionable assumptions. Because it became the 'gold standard' of flood system performance, the SPH served to prevent up-to-date re-analysis of the true risks of catastrophic flooding."

The MR&T levees are designed to hold back a mythical super-flood, while the New Orleans system was designed to defend against a category 3 storm at best, but the problem with the methodology remains the same. The deterministic approach doesn't plan for the system to fail, nor does it compute uncertainty, or risk, so the assumptions it makes are never as informed as those incorporated into a probabilistic analysis.

In the late 1990s, various experts and agencies, including the Corps, began saying publicly that the levees around New Orleans (notwithstanding design flaws later revealed) could not handle a big hurricane. The Corps considered the flooding of the city so likely that it devised a detailed plan for "unwatering" the metropolitan area. In 1999 Congress authorized the Corps to study various proposals for protecting Southeast Louisiana from a category 4 or 5 storm. Government scientists began running simulations on a supercomputer, analyzing what a big hurricane could do to the region. The agency needed $8 million and six years to finish the study.

"That's a lot of time," said an engineer working on the study in 2003. "Hopefully we won't have a major storm before then."

In July of 2004 FEMA performed a separate exercise, coordinating with officials from fifty state, federal, and volunteer organizations to anticipate the impacts of a high-end category 3 storm. The hypothetical storm was called Hurricane Pam and its hypothetical effects were terrifying. With sustained winds of 120 miles per hour and up to twenty inches of rain, Pam's storm surge overtopped levees, forcing more than 1 million residents to evacuate and destroying at least five hundred thousand buildings. The exercise predicted with prescient accuracy what would happen when the real thing spun into the Gulf a little more than a year later.

On the morning of August 29, 2005, Katrina smashed into Southeast Louisiana's outdated defenses. Floodwalls blew out before they should have, some were overtopped by a storm surge they weren't designed to handle, others hadn't been maintained by their local sponsors. Nearly two thousand people died, damages exceeded $100 billion, and 85 percent of the New Orleans metro area was flooded. The Corps took the blame.

Stinging from the bad press and public outcry following the disaster of Katrina, the Corps moved quickly to build a new flood protection system for New Orleans. The $14.6 billion network of levees, pumps, and gates was eventually named the Hurricane Storm Damage and Risk Reduction System, the HSDRRS. (After Katrina, the Corps was careful never to promise "flood control" or "flood protection." All definitive terms were replaced with legalistic vagaries. Levees, no matter how well built, now only provided "risk reduction." This term was, of course, more accurate, but people in the Mississippi Valley felt betrayed; the Corps had been telling them for a hundred years that it did control floods.)

Tom Holden Jr. was the top civilian at the New Orleans District when the HSDRRS was designed. Of the deterministic, or project hurricane approach, Holden said, "That's the way we used to do it, and then Katrina came." In the Corps' final report on Katrina, the first "lesson learned" was that the standard project hurricane methodology was "outdated and should no longer be used. More flexible and robust probability-based methods are available." For the HSDRRS, Holden said, they used a probabilistic model. They also added features that would allow the levees to survive overtopping if a higher storm surge should come along. Another hard-learned lesson: During Katrina, over-topped levees eroded below sea level, and drying out the city took weeks.

Even with the best engineering expertise and a tremendous supply of money and political will, the HSDRRS structures were sinking faster than expected. The concrete was barely dry before the Corps announced that within four years the levees might no longer qualify as one-hundred-year protection. "Absent future levee lifts to offset consolidation, settlement, subsidence, and sea level rise," stated a government press release, "risk to life and property in the Greater New Orleans area will progressively increase."

Katrina was considered a four-hundred-year storm on the Mississippi coast, a two-hundred-fifty-year at New Orleans East, and a one-hundred-fifty-year on the city's north shore. The HSDRRS—at best—protected against a one-hundred-year storm, an event with a 1-in-100 chance of occurring in any given year. That's how much risk reduction $14 billion bought for New Orleans. Congress didn't want to pay for more.

At least for the HSDRRS—limited though its protection might be—the Corps had thoroughly assessed the risks and designed the system accordingly. Better science and technology could also improve the MR&T levees, yet it wasn't being done. Would it take a Katrina-style catastrophe on the Mississippi to get the Corps to abandon its deterministic model?

"I don't think today we're as comfortable as we were in '73," said Holden, referring to the last time the Corps revised the project flood flowlines that dictate levee height. "You should really have a probabilistic-based event," he said. There does seem to be more rain and more flooding. "So what's different and what do you do about it?" It's not really up to the Corps to sound alarm bells or propose rebuilding the entire flood protection system for the Lower Mississippi, said Holden. Such radical changes would have to come from Congress.

"Right now, I think, as a country, we've got our heads so far up our ass on issues that aren't the most pressing issues," Holden said. "It'd be wonderful if we could have frickin' power that uses no fossil fuel; everybody had frickin' healthcare and it was free; nobody had to take out a loan for college—but this shit ain't real." What is very real to Holden is the threat posed by the Mississippi River.

"In my view, we do not have a Congress that's trying to solve problems," he said. The question of preparedness on the Mississippi River and vulnerability to super-floods, he said, "is a big problem." A levee breach between Baton Rouge and New Orleans could easily cause a recession by destroying a substantial percentage of the country's oil refining and chemical production as well as most of its grain exporting capacity. "This needs to be attended to preemptively and decisively," Holden said. Banking isn't the only system that's too big to fail. The country can't afford to stand by while its flood defenses crumble.

Congress first needs to recognize the problem. Then it needs to ask the Corps how to solve it. The Corps could then recommend a probabilistic analysis instead of a deterministic one. Congress could authorize the Corps to perform a study, perhaps called "Continuing the American Way of Life in a Wetter World." With such a proposal in hand, said Holden, all Congress would then need is "the political courage to do something."

Louisiana has marshaled much more courage than the federal government. It has had no other choice. After Hurricane Katrina, the state combined coastal restoration and flood protection into a new single agency, the Coastal Protection and Restoration Authority (CPRA). In 2017 CPRA released a fifty-year, $50 billion Master Plan for defending the coast from hurricanes, sinking land, and rising seas. The HSDRRS was not nearly enough. More had to be done, and fast. The state had already built thousands of acres of artificial islands and marsh, and hundreds of miles of levees and seawalls. Even so, the Master Plan revealed, if it stopped there, 2,250 square miles of additional land could be lost and flood-related damage could top $150 billion.

Engineering works, no matter how huge or expensive, were not enough to protect the coast. Louisiana sought to marshal a much more powerful force: the Mississippi River itself.

PART III

Rivers
of Earth

The captain navigated not by what he saw in front of him, but by using a map in his mind. He had been plying these waters all his life, and in that time the landscape had changed almost beyond recognition. Greg Stockholm, a sixty-one-year-old auto body repair specialist from Springfield, South Dakota, used to water-ski from Springfield's boat ramp clear across the expanse of Lewis and Clark Lake to the bluffs on the Nebraska shore. Back then, the only thing that might have caught his boat's propeller were trees, buried underwater when this part of the Missouri River Valley was flooded to create a reservoir. Now, barely a hundred yards out from the boat ramp, a water-skier would plow smack into a grassy island.

It was the Fourth of July, and Stockholm was taking his wife, Sandy, and three buddies on a tour of the former lake. "About mid-seventies, started running into sandbars," he said, at the helm of his thirty-foot express cruiser. The boat—white fiberglass with teal trim—had berths for six passengers but could float in just twenty-two inches of water. That number mattered because the tangle of channels it negotiated was getting shallower.

After a few turns into the marshy maze, the boat ramp disappeared from view. Stockholm soon saw signs of shallow water all around. He hit a switch to raise the propeller. "No water there," he said. "We're gonna try a different route." A loud beep drowned him out—the cruiser was warning him it was about to hit bottom. The engine shut off automatically. Stockholm restarted, and throttled into reverse, churning up clouds of mud.

Gazing across the strands of river, he distinguished ripples caused by the wind from ripples caused by sandbars. He maneuvered upstream, skirting the fringed ends of some bars, running aground on others. Stockholm had navigated these waters almost daily as a young man, and regularly until his mid-fifties, but he hadn't been out for a year, and the landscape was changing so fast, he didn't know it anymore. The depth-finder's screen flashed erratically—three feet, nineteen feet.

Stockholm knew the water would be deep beneath an earthen bluff on the Nebraska shore. He looked for landmarks to line up on the way back. "I'll take that point and look back there at those trees and take that as a target," he said. He might as well have been quoting Mark Twain, who, as a cub pilot learning the Mississippi in the 1850s, was told that he needed to know the river like the hallway in his own home, at night, with no candle. And the hallway would change continually. He'd have to recognize when the river was building a bar or eroding one, cutting a bank or extending one. In *Life on the Mississippi*, Twain recounted this exchange with Horace Bixby, the pilot who had agreed to "learn him the river":

> "Do you see that long slanting line on the face of the water? Now, that's a reef. Moreover, it's a bluff reef. There is a solid sand-bar under it that is nearly as straight up and down as the side of a house. There is plenty of water close up to it, but mighty little on top of it. If you were to hit it you would knock the boat's brains out. Do you see where the line fringes out at the upper end and begins to fade away?"
>
> "Yes, sir."
>
> "Well, that is a low place; that is the head of the reef. You can climb over there, and not hurt anything. Cross over, now, and follow along close under the reef—easy water there—not much current."
>
> I followed the reef along till I approached the fringed end. Then Mr. Bixby said—
>
> "Now get ready. Wait till I give the word. She won't want to mount the reef; a boat hates shoal water. Stand by—wait—WAIT— keep her well in hand. NOW cramp her down! Snatch her! snatch her!"

Stockholm could have given identical instructions to someone learning to navigate the headwaters of Lewis and Clark Lake. These waters

changed so often that no one had charted them. None of these islands
or channels even had names.

Inquiring after an unfamiliar tributary of the Mississippi, seventeenth-
century French explorers recorded an Algonquian word that meant
"muddy water," a fact William Clark knew from unpalatable experience.
In his journal of June 21, 1804, Clark wrote, "The water we Drink, or
the Common water of the missourie at this time, contains half a Comn
Wine Glass of ooze or mud to every pint." When Clark recorded those
words, his expedition was approaching the site of modern-day Kansas
City, and the ooze they didn't swallow was flowing down the Missouri,
into the Mississippi, and on toward the Gulf of Mexico. Half the Missouri's watershed is semi-arid, and although its basin stretches from the
Rockies to St. Louis, and from Saskatchewan to Kansas, the river carries
comparatively little water. What it does carry is sediment. At least it
used to. In pre-dam times, the Missouri supplied more than half the
Mississippi's sand, silt, and clay. Now, thanks to dams on the Missouri
and Arkansas Rivers, the Mississippi's sediment load has declined by
almost two-thirds.

Gavins Point Dam is the smallest and southernmost of the six dams
that slow down and back up an eight-hundred-mile stretch of the Missouri River. Spanning the river near Yankton, South Dakota, it consists
of a mile-and-a-half-long grass-covered slope, a powerhouse with three
turbines, and a concrete spillway topped with fourteen Tainter gates.
Gavins turned a shallow, fast-moving river into Lewis and Clark Lake,
a twenty-five-mile-long reservoir that flooded the valley from bluff to
bluff. As soon as Gavins began holding back water, in 1955, the newly
formed lake began filling with sediment.

Another river also disgorged into this lake—the Niobrara, which
drains some of the driest land on the Great Plains, including the Nebraska
Sand Hills. It, too, carries a lot of sediment. Before the dam was built,
a small delta of pure sand would form where the Niobrara flowed into
the Missouri. When the Missouri flooded, the delta was washed away.
But sand is heavy. To carry it, water must move swiftly. When Gavins
Point was completed, the Niobrara was suddenly dumping sand into

a still lake instead of a fast river. Because other dams were operating upstream, the Missouri lacked the force to scour the sand away. It piled up in the riverbed and stayed. Deltas form where rivers hit still bodies of water—usually oceans, but in this case, a man-made lake.

The Missouri dropped its own heavy sediments at the same spot, and, as this mass of material accumulated, it advanced into the reservoir. It broke the surface. Islands formed. Trees and grass took root. The delta expanded and the lake contracted. By 2018, only fifteen miles of the original twenty-five-mile reservoir could be called a true, open-water lake. The rest was a textbook delta, a maze of tentacular channels and hummocky shoals. Except for the bluffs, it could pass for the mouth of the Mississippi, the Danube, the Ganges, or the Mekong.

As the Lewis and Clark delta grew larger, a useful reservoir grew proportionately useless. For lack of a better word, specialists in the still-new field of sediment management declared Lewis and Clark Lake 30 percent "full" in 2018. Full of sediment, not water. Any dam that isn't moveable like Olmsted must gradually surrender its capacity for storing water to the relentless assault of sediment. The reservoirs behind all seventy-nine thousand dams in the United States are filling at different rates. Reservoirs in the arid American West, where man-made systems are often the only source of water, are filling the fastest. Thirsty for the power and wealth to be gained from controlling these rivers, American engineers and politicians built thousands of dams in the first half of the twentieth century. None were built with a way to manage sediment.

Greg Stockholm had changed out of his work clothes and put on a Hawaiian shirt and white jeans before trailering his express cruiser down to the boat ramp. He knew the delta by feel, by instinct. His wife, Sandy, knew it in terms of project beneficiaries, operations and maintenance, local sponsors, and continuing authorities. She knew the difference between Section 1179 and Section 1119 of Public Law 114-0002.

As the executive director and the only paid employee of the Missouri Sediment Action Coalition, Sandy Stockholm was engaged in battle, as a video on her website put it, against the "creeping cancer that is sediment in our reservoirs." Her coalition originally aspired to be a lobby, but couldn't raise significant money, and so became a nonprofit organization, attracting about $25,000 a year in membership dues, plus the occasional grant. The funding covered Sandy's part-time salary and her travels around the region, promoting research, hosting meetings, and setting up booths at fairs—all in the interest of convincing small-town Midwesterners to care about sediment. It's a passion she shares with a man named Paul Boyd. In matters of sediment—and the ambition to navigate around a sedentary bureaucracy—they spoke the same language.

Boyd, an energetic forty-four-year-old who works out of the Omaha Federal Building, is one of the U.S. Army Corps of Engineers' principal sediment experts. He was an agricultural engineer by training, yet as he studied the way soil eroded, moving off farmland and into waterways, he began to wonder where it was all going. Officially, Boyd's title is hydraulic engineer. "Sediment engineer" isn't a formal job title at the Corps—nor does it exist as a college degree.

The Corps is devoted to water management, said Boyd, "but they're managing half the equation." In Boyd's opinion, the Corps knows a lot about how water moves and is able to control it fairly well. It knows far less about sediment—the bits of earth and rock held in suspension and then set down by moving water. The Corps has made many consequential and seemingly irreversible decisions, such as leveeing off the Mississippi River before and after the 1927 flood, building dams across its tributaries, and cutting off its bends—without adequately considering sediment.

Like the iceberg that doomed the *Titanic*, the slow buildup of sediment in reservoirs hadn't yet alarmed the Corps, said Boyd, because the ship was still afloat, the reservoirs still functioned. The Corps' economic model, which determined what the agency studied and spent money on, wouldn't acknowledge an incremental problem like sedimentation until it reached a crisis point—say, until Lewis and Clark Lake became so clogged that it stopped functioning, in about one hundred years. Though it could see the collision ahead, the Corps wasn't altering course.

"What is the benefit for your grandchildren—to have reliable hydropower from the lake, or to have flood-risk reduction from the lake?" Boyd said. "What is the value to them in the future, and what is it worth? What should we spend to preserve that future?"

The government hadn't given him the time or the money to find out. According to Boyd, the best work on the subject had been done by Rollin Hotchkiss, a professor of civil and environmental engineering at Brigham Young University. Through the Freedom of Information Act, Hotchkiss had acquired reams of material from Boyd's office, and in 2015 had published a paper that laid out a case against the Corps' budgetary myopia.

When deciding whether or not to build a project, Hotchkiss found, the Corps looked at benefits and costs, near-term and long-term. The near-term benefits of a typical dam were flood control and water storage. In 1955, a dam that did only those two things would have cost, say, $50 million; in a hundred years, such a dam might fill with sediment and be useless. By contrast, $55 million in 1955 might have bought a dam that managed sediment, and so would continue to control floods and store water for a thousand years. But by the Corps' calculations, it wasn't worth spending that extra $5 million in 1955 to keep a dam operating beyond 2055. The Corps favored investments with immediate returns over projects that wouldn't deliver their benefits for fifty or one hundred years. Benefits more than thirty years in the future, Hotchkiss discovered, were considered essentially worthless by the Corps.

By this logic, it didn't pay for the government to counteract slow-moving threats like sedimentation. Or nuclear waste. Or climate change, wrote Hotchkiss. Because they didn't merit preventive measures today, these problems—incrementally bad now, potentially catastrophic later— fell to future generations. Sandy Stockholm and Paul Boyd spoke of "intergenerational equity," the idea that it was unfair for those of us alive now to reap short-term gains when our kids, and our kids' kids, would suffer the consequences.

"Some folks would say, 'Well, we don't pay to manage sediment,' but you can make the case that we pay *not* to manage sediment," said Boyd. Gavins Point Dam did cost $50 million to build in the 1950s, and it was not built with a way to manage sediment. Hotchkiss estimated that taxpayers had spent, or would soon spend, at least $258 million to address the effects of sedimentation around Lewis and Clark Lake. They'd paid to move water intakes and boat ramps; to buy out thousands of acres of property; to move roads, neighborhoods, and whole towns. And such costs would only increase—and Gavins Point was just one dam.

Boyd was delighted that Hotchkiss had made these figures public. Privately, he suggested a total even higher than Hotchkiss's estimate. The cost of not managing sediment at Lewis and Clark Lake was fast approaching $300 million, the low end of Boyd's estimate for retrofitting the dam so it could manage sediment. These modifications wouldn't flush out much of the muck that had already accumulated in Lewis and Clark, but at least they would stop it from getting any worse.

Every river has a sediment diet, the amount of silt, clay, and sand that its water naturally carries. The Missouri was starving; its dams were capturing sediment and not letting it go. Unless the river was in flood, all the water escaping from Gavins Point Dam passed through the powerhouse, where intakes had been deliberately positioned to avoid sediment-laden water. (Turbines that are spun by moving water to generate electricity get ground down and gummed up if sand mixes with that water.) Leaving Gavins Point, the Missouri River is basically clear. In the language of hydrology, water without sediment is "hungry." Hungry water eats downward, deepening a river's channel, and outward, slicing away a river's banks. "The river gets what the river wants," Boyd said. At Yankton, the town immediately downstream from Gavins, the Missouri's bed has sunk in some places by eleven feet. The river's banks are now dry ledges high above the sandbars. Young cottonwoods won't grow because their roots can't reach the river. Red cedars, invasive upland trees that don't tolerate water, are taking over.

The Missouri is hungry from Gavins Point to the mouth of the Platte River, south of Omaha. The minimally dammed Platte nourishes the Missouri for a few hundred miles but, by the time it reaches Kansas City, the river is hungry again. Here, too, the Missouri's bed has dropped as much as ten feet. In 2017 the Corps' Kansas City District reported that, if nothing was done to adjust the river's diet, the bed could fall five additional feet. The cost to upgrade or move all the infrastructure affected was estimated at $785 million—plus $29 million a year in increased operations and maintenance costs—payable by future generations.

Those figures didn't include money already spent to shore up gnawed-away bridge piers, dump rock on eroding levee toes, and move water intakes. In 2003, the Missouri dropped below the intakes of WaterOne, which serves 430,000 thirsty customers near Kansas City. The utility had no choice but to float a barge out into the channel and pump water from there. The following year, WaterOne spent more than $2 million of its own money to build an auxiliary low-water intake.

Engineers knew that to satisfy the Missouri's hunger they had to feed it. If its dams could begin to pass sediment, it would quit gorging on its own bed. In the public mind, however, sediment's name was mud. It makes water cloudy, and cloudy water looks dirty.

The Clean Water Act of 1972 officially listed sediment as a pollutant. Thirty milligrams per liter, the limit under the act, Boyd said, was "essentially clear water. It's not muddy water, or slightly muddy water." When the act was passed, most of the country's dams were operational and the water pouring over their spillways and through their turbines was unnaturally clear. Before logging and agriculture, many smaller American rivers did run clear, or close to it. But, even in their pristine state, the big Western rivers ran like chocolate milk. Yet, in 1972, clear water was decreed clean, and muddy water polluted. The act implied that all American rivers should be cleansed of sediment, when, in fact, clear water was "an artificially manufactured result of a lot of infrastructure," Boyd said. The act made it all the harder for him to convince people that the one thing the Missouri needed was more sediment.

Sediment is not sexy, explained Sandy Stockholm, Boyd's steadfast pro-mud proselytizer. Driving along a lonely South Dakota highway one day, she reflected that "getting people excited about this issue is like, if we're driving along the road and there's a bunch of potholes. Sure, people are concerned about it, but they think someone's going to take care of it. They're not going to form a group and go to a meeting every month and really get active about it." It's also hard to get people excited in a place with so few people. The region straddles a borderland between the Great Plains to the west and the lush farm country to the east. Trees grow smaller and farther apart; cattle range begins to take over from row crops; settlements and buildings become scarce. Yankton, once the territorial capital, is the biggest city around, with almost fifteen thousand people. Named for a Sioux tribe exiled farther upriver, it is an outpost on the edge of a vast emptiness.

Gavins Point Dam was plunked down four miles west of Yankton for reasons of politics, not continental water management. No clear-thinking engineer would have put a dam there, downstream from the Niobrara-Missouri confluence. A dam above the confluence would have made real sense. That dam could have held back the Missouri before the sediment-laden Niobrara emptied into it. All that Nebraska sand would then have been carried down toward Omaha and the Gulf. But regional politicians—with assistance from the Corps—convinced themselves that the outflows from Fort Randall Dam would scour the troublesome delta away. Building the dam close to Yankton meant that it could hold back more water and, according to the Corps' 1949 study of the project, "the recreation opportunities for the people residing in the bordering states of Nebraska and South Dakota would be tremendous."

"Politics were part of the process from the beginning," Sandy Stockholm said. Politics won out over engineering. Yankton got its lake.

Greg Stockholm's father was born in the village of Niobrara, Nebraska, slightly east of the Niobrara-Missouri confluence, on the Missouri's south bank. When his father was a kid, Greg said, the village stood on a bluff at least twenty feet above the Missouri. But as Gavins Point went to work, trapping the Niobrara's sand in the newly formed Lewis and Clark Lake, the Missouri's high-water mark crept steadily up the bluff. Imagine getting into a bathtub and lying down. The water in the tub will rise. As sediment settled in the lake, the lake rose the same way.

By 1971, groundwater was seeping into Niobrara's basements. The surrounding farmland was too soggy to support a tractor. Corn didn't grow. That year, an Omaha law firm filed four suits against the Army Corps of Engineers on behalf of farmers whose crops were gagging on mud. The cases were consolidated, and Rick Spellman, an attorney with the firm, argued *Barnes v. United States* in the U.S. Court of Claims. Though the village of Niobrara wasn't party to the suit, a U.S. senator from Nebraska noticed that the village, too, was in need of aid. A Court of Claims judge ruled in 1976 that the Corps was to blame for clogging the Missouri with sediment that should have been on its way to Louisiana. The government was compelled to buy about $2 million worth of flowage easements on thousands of acres of farmland in Nebraska and South Dakota. Spellman's firm also helped manage $18 million that Congress had appropriated to move the village.

Spellman spent years working on the case. He made lifelong friends and eventually bought two houses in a riverside subdivision. Most weekends, he drove the three hours from Omaha with his wife, son, and two daughters. These days, his adult children bring the grandkids. Spellman also became an active member of Sandy Stockholm's Missouri Sediment Action Coalition.

Now seventy-five years old, trim and tan, Spellman had driven up on this Fourth of July weekend in his Jeep SUV. He was giving a tour of Niobrara's faint traces: "That's the old main street. See how wide it is?" The broad ribbon of concrete led nowhere. "This'd be the town center." The brick storefronts and frame houses had vanished, along with the town's grandest building, the Masonic Temple, built of stone in 1914. Now the site was flat and featureless; where six hundred people once lived, the Corps had built a municipal golf course.

This abandoned settlement was, in fact, Niobrara's second location; the town had moved once before. In 1881, twenty-five years after Niobrara was founded, a winter flood destroyed the original town. Sheets of ice sliced into its buildings, and those left standing filled with freezing water. The townspeople decided to move. In a history subtitled, "The Town Too Tough to Stay Put!" John E. Carter wrote: "Teamsters, armed with house jacks, winches and capstans, block and tackles, beams, poles, oxen, mules, and horses began raising, bracing, and hauling building after building to the new Niobrara townsite." The movers rolled buildings on wagon wheels up to "the bench," a mile and a half to the southwest, where they had mapped out the new town. By early 1882, most of the houses had been relocated—along with two newspaper offices, three general stores, two drugstores, two hardware stores, a harness shop, two blacksmiths, five hotels, two livery stables, three doctor's offices, a schoolhouse, and a church.

Niobrara's second home atop the bench was higher and less flood prone. If the Missouri had never been dammed, the resettled town could have weathered just about anything, but community leaders hadn't anticipated the machinations of the Army Corps of Engineers.

No buildings were rolled to new sites in the 1970s. Everything but the school was demolished. A new town was platted on the bluff above the flood zone. The town organized itself around a business district on both sides of the rerouted Highway 12. Townspeople bought lots with their condemnation proceeds and whatever other money they had. Home sites were awarded by raffle, but a secondary market soon arose. People traded and traded again until they had reestablished familiar enclaves and moved next door to old neighbors. Spellman said that "a sociologist would have a field day" analyzing the social dynamics.

Although the community chose to name its seventeen streets after trees, trees did not naturally grow on the bluff. The fine sandy soil, dug up for foundations and yards, would blow across town and stick to the freshly painted houses. Transplanted seedlings eventually took root and the sand stopped blowing, but more than forty years later, the new Niobrara still felt incomplete. Nearly all of its businesses operated out of one long, sheet-metal shed called "the mall." The school remained below the bluff, at the old town's highest point. If the Niobrara River ever veered from its current bed, its most likely trajectory would cut through the school's playground, a possibility that punctuated Niobrara's bleakest fact: The town had lost more than half its population. It had 736 residents before the second move; the latest census estimate was 317.

At the heart of the growing Niobrara delta, an abandoned railroad bridge crossed the river's mouth. The main channel ran swiftly along the Niobrara's west bank; a minor channel ran on the east bank. Between the two channels, a huge sandbar had grown up under the bridge. The old state park lay just upstream; it had attracted more than a hundred thousand visitors a year, before it melted into an indistinct tangle of vegetation. Legend has it that a pilot once flew a small airplane under the bridge, but on this Fourth of July, only ten inches separated the bar from the rail bed. Beyond the bridge, lobes of sand and marsh extended out of sight, braiding the delta of Lewis and Clark Lake.

Acreage that was damp in 1971, when Spellman filed suit against the Corps, was underwater now. Protruding pilings or a few boards nailed together revealed once productive property. Streaks of water resembling canals were actually submerged roads. Driving east on Highway 12, Spellman gazed out at marsh grass and swaths of open water. The grass, blackened by the sun, smelled like rotten eggs. Remembering the late 1960s, he said, looking left, "Good farmland." He nodded right, "Great farmland." He added, "It's an unintended consequence of doing something they shouldn't have done in the first place without planning ahead for it." This stretch of Highway 12 and another west of town ran only two feet above the former farmland. The federal government was under obligation to raise it. The most recent estimate for the job was $160 million.

In a few miles more, Spellman came to Bazile Creek and the edge of the Santee Sioux reservation—173 square miles of crop and rangeland, which became home to the tribe in 1866. Dead cottonwoods stood, black and spiky, at the creek's mouth. Cottonwoods love water, but too much of it can kill them. Here, they had drowned.

The Santee Tribe was living in Southern Minnesota when Europeans first encountered it. By 1862 the U.S. government had restricted the Santee to a reservation along the Minnesota River. There they struggled with poor harvests and scarce game. Government subsidies—which had been promised by treaty—lagged, and traders refused credit. The tribe was starving. "When men are hungry, they help themselves," the Santee leader, Little Crow, wrote to a government Indian agent. Returning from a hunt, four Santee men came upon the farm of a white settler. What began as a dare to steal some of the farmer's eggs escalated, and the farmer and four other settlers were killed. Knowing that the murders wouldn't go unpunished, several Santee bands saw an opportunity to repel the invading settlers, whose government was occupied with the Civil War. The Santee started a war of their own, attacking farms and villages. Before the fighting ended, more than six hundred whites and as many as one hundred Santee had died. The courts found 303 Santee guilty of murder and sentenced them to death. President Abraham Lincoln pardoned all but 38. These 38 men were hanged in 1862 in the largest mass execution in U.S. history.

The remaining sixteen hundred Santee, mostly women and children, were interned for the winter on the flats at the confluence of the

Minnesota and Mississippi Rivers. Cold and hungry, hundreds died. The following spring, Minnesota evicted the tribe from the state. The surviving Santee were shipped by steamboat, first to a reservation farther up the Missouri, where they lived as prisoners and starved for three more years, and finally to the mouth of the Niobrara.

The government wanted the Santee to become self-sufficient small farmers, true to the Jeffersonian ideal. Some, like Mike Crosley, the Santee's land manager, did become successful farmers—at least for a time.

Crosley's family had lived on the reservation since the beginning. "Got the old stories of when the dam was built—when they moved everybody off what we called the river bottom," he said. Crosley's dad owned the International Harvester dealership in Niobrara. He sold it when the town was relocated, and he moved his family to Santee, the reservation's village, ten miles downstream on the Missouri. "My great-grandfather moved off the river," Crosley said. "My grandfather and uncle moved off the river. My dad moved off the river. We're used to it. Now I'm moving off." He was planning to build a new house on higher ground, "where I won't live long enough to have to move again."

His current home sat just past the mouth of Bazile Creek off Highway 12. The tribal office, where Crosley worked, was on the hill above his house. From his office window, he could look across the creek at the land he and his father used to farm. He couldn't farm that land anymore because it wasn't land anymore. In thirty years, the creek had risen ten feet—not during floods or wet seasons, but permanently. Back when the ground was dry, Crosley grew corn. As sediment filled the lake and the water table rose, he tried soybeans, then alfalfa, then hay, the most water-tolerant crop he could think of. Then he gave up. The ground turned muddy, then boggy. Now it grew cattails, reeds, and canary grass.

"You planned a future on that land," Crosley said. "Your culture is out there, your family is surviving and going around time after time." He had sued the government for all the damage done, but no single payment, no matter how high, Crosley felt, could compensate the future generations of Santee Sioux who might have made their livings from that land.

The Fifth Amendment to the United States Constitution says, in part, "nor shall private property be taken for public use, without just compensation." Known as "the takings clause," it means the government cannot seize private property—exercise its right of eminent domain—without paying for it. In 1976, Rick Spellman argued in front of a federal judge that the Army Corps of Engineers "took" the farmland surrounding Niobrara: The Corps built a dam; the dam caused a sediment buildup. The emerging delta backed up the river and made the water table rise, turning farmland to bog. It all added up to a "taking," Spellman contended, and because the Corps hadn't paid for what it took, it violated the Fifth Amendment.

Spellman's case, *Barnes v. United States*, set a trend, if not an absolute precedent. "Takings" claims became a regular tactic among property owners seeking to hold the Corps responsible for flooding. Farmers in southeastern Missouri filed a takings case over the 2011 activation of the Birds Point–New Madrid Floodway. (They lost.) St. Bernard Parish, southeast of New Orleans, filed a takings case after its levees breached during Katrina. (It lost.) Farmers in Iowa and Missouri sued the Corps in 2014 for "allowing" the Missouri River to flood so often and for so long that it amounted to a seizure of their fields. (So far, they've mostly won.) In 2019, the State of Mississippi sued over land that could no longer be farmed because, it claimed, the Corps wasn't diverting enough water from the Mississippi River into the Atchafalaya.

Mike Crosley was in the process of settling his third takings claim, this time for his home and twenty acres of land. The Corps had agreed to settle, but Crosley felt cheated. "We're not even getting a third of the value," he said—$30,000. "How do I pack up that house? Where do I go? This is where I live. Eleven years, I've been fighting with them, and so I haven't put any improvements in the place. Why do you want to pour money down a rat hole?"

Because he owned his land privately, Crosley had been able to make a direct claim against the Corps. The Santee Sioux Nation couldn't do this. At least a thousand acres of the reservation had been waterlogged, Crosley figured. The Bureau of Indian Affairs administers tribal lands, and it should have taken the Corps to court on the Santees' behalf. But it hadn't. The Nation would have to sue the Bureau to make that happen, Crosley said wearily. A suit to force a suit was too disheartening to contemplate. It wasn't until 2005 that they were paid, in a separate claim, for the taking of land flooded by Gavins Point Dam fifty years earlier. And that money wasn't theirs to spend. They received only the annual interest payments. They couldn't touch the principal.

Crosley was discussing this one day with the tribe's chairman, Roger Trudell, during an informal meeting at the tribal headquarters on top of the bluff in the village of Santee. Trudell had been up to his neck in sediment for a long time. He recalled meeting an Army Corps colonel to talk about it back in the 1980s. The colonel admitted that sediment was a problem, but didn't say he'd do anything about it. Every time tribal leaders went to Washington, they talked about sediment, Trudell said, "but we've never had a positive response."

Trudell chuckled, a sound not mirthful but tired and resigned.

"Can't stop siltation," he said. "There's no way to stop it."

What about calling Nebraska's delegation to Congress? The state had two senators and three representatives.

Trudell chuckled again. "I don't even think they know who we are."

But he had another reason for speaking softly on watery matters. Like other tribal nations, the Santee had a right to all the water they would ever need. In a fight with the Corps over sediment, they worried, a court might force them to put a number on that amount—to "quantify" their water rights. Some tribes had done so and received huge allocations, but the Santee refused on principle.

"It belongs to us. We don't need to quantify something that's already ours," Trudell said.

The Santee could make a strong case that sedimentation was fouling their drinking water. The whole reservation drew upon three shallow wells near Bazile Creek, each no more than twenty-five feet deep. The water had to travel through nine miles of pipes to reach the village. Unfiltered, it was foul. "You have to run it through something," Crosley said. "Otherwise it's the nastiest water that you ever washed your clothes in." Yet, more than a thousand people depended on it. Now, with both the creek and Lewis and Clark Lake rising, even nastier water was inching toward the tribe's wells.

The Santee badly needed a better source of water. But a design for a new one—drawing from Lewis and Clark Lake, for instance—would involve the federal government, which would probably require the tribe to violate its principles and quantify its water rights.

"We need a new water system," Trudell said with finality. "The real challenge is how are we going to stop losing land. I have no idea how we would do that."

Sandy Stockholm—always looking for stakeholders to engage with—had sent Mark Simpson, the Missouri Sediment Action Coalition's board president, to this meeting with Crosley and Trudell. Simpson told Trudell that the Corps had been authorized to study how to potentially remove sediment from the lake. However, the government wouldn't begin until it found a local sponsor to cover half the study's cost, about $200,000.

The coalition, Simpson said, wanted to try out a sediment-collecting device developed by an Ohio company. The collector was a hump-shaped tank with a slot on top, placed like a speed bump on a stream's bed. As water coursed over it, sand fell through the slot, and an auger pumped it onto dry land. The Corps had a small collector of its own and might be willing to put it to the test on the Niobrara, Simpson said. Once collected, the sand could be studied for potentially profitable uses, like mixing it with melted plastic bottles to make asphalt, or adding it to cement for building blocks. But Simpson couldn't deny reality. Any industry that needed sand had countless other sources. The cost of hauling sand from the remote Niobrara quashed any hope of profit.

"This green that you're seeing out here," Simpson said, waving a hand toward the delta outside, "there's probably never going to be any change in that. What we can do is try and divert it from getting worse than it already is." He hoped that someone from the tribe would come to the coalition's next meeting; maybe the Santee Sioux Nation would be interested in helping fund the study.

South Dakotans should pay for it, since they're the only ones who benefit from the dam, said Trudell.

"It's a political thing. For Yankton to have recreation is really what that dam's purpose is today," Crosley had said earlier. "As the siltation problem continues and there's no capacity out here, then what's the purpose of that dam?"

While the Corps might not admit to building Gavins Point Dam for Yankton's recreational amusement, it was never intended to control floods. When Fort Randall Dam, sixty miles upstream, had to release water, Lewis and Clark Lake wasn't big enough to contain the Missouri River for long. It was the job of four dams farther upstream—Big Bend, Fort Peck, Garrison, and Oahe—to protect Omaha and Kansas City. The huge reservoirs those dams created were practically immune to sediment; they wouldn't fill with it for a thousand years. Gavins Point's primary function was to calm waters irregularly released from Fort Randall's hydropower plant. Gavins was a "reregulation dam," built solely to smooth the passage of barges navigating downstream.

But what barges? By one estimate, the annual benefit of river traffic in the entire Missouri basin came to $7 million. No study had ever measured that benefit against the cost of maintaining the river's nine-foot-deep, three-hundred-foot-wide, six-hundred-mile-long navigation channel for nine months of the year. (One was begun but never completed.) The Corps had also spent $800 million to buy land and to build habitats along the Missouri for two species of birds and one species of fish that its dams and channels had endangered. If the Corps gave up the navigation channel, it could stop building habitats. Gavins could simply release water to mimic the river's natural highs and lows. In fact, by not catering to barge traffic, there would be no need for reregulation—and no need for Gavins at all. What costs and benefits would result from dismantling the dam?

No one had done the math. Though the barges amounted to little economic importance, their right to float on the river was fiercely defended. The pressure to maintain the navigation channel, some in the basin surmised, came not from shippers but from farmers. Perhaps the most powerful constituency in the region, farmers did not want to see the Corps playing around with the river. Both shippers and farmers wanted a predictable river, while birds and fish and plants thrived on variability. The farmers wanted flood control above all, and the Corps had seven more "above alls" to juggle on the Missouri by congressional mandate: fish and wildlife, water supply, water quality, irrigation, recreation, hydropower—and, of course, navigation. No wonder that arguments over the Corps' priorities here often went around in circles.

Sediment management wasn't even on the list.

The meeting at the Santee Sioux headquarters had come to an end. As Mark Simpson was leaving, chairman Trudell told him, "If you don't like sediment, I'm with you."

On the wall behind Trudell hung an old photo of the wood frame church that commands the high ground on the village's outskirts. In the picture, the church's brilliant white paint stood out against a sparkling blue lake. Simpson had driven past the church earlier that day. The white paint still gleamed, but the background didn't sparkle. It was dull green.

"It used to be beautiful coming over the hill," Trudell said. He could remember swimming in that blue, blue water.

*   *   *

The Mississippi River flood of 2018 was near cresting as Major General Richard Kaiser strode energetically across the top of the Low Sill Structure, an unremarkable row of battleship-gray gates lifted by a boxy crane. The Low Sill is part of Old River Control, a complex of dam-like barriers south of Vidalia, Louisiana, designed to prevent the Mississippi from changing course. As president of the Mississippi River Commission, Kaiser was responsible for keeping North America's greatest river under control. With a tearing sound, a whirlpool opened beside the structure, an intermittent effect of the huge volumes of water passing beneath it. The void, spinning and warping downward beneath the river's surface, was the width and height of a full-grown person. "Thought I saw Jimmy Hoffa in there," said Kaiser.

The modern Mississippi flows past Baton Rouge and New Orleans and out to the Gulf through a fork of channels called the Birdsfoot. For nearly a thousand years, the river has deposited sediment along this route, making it longer and less steep. Like every drop of water, the Mississippi wants the shortest and steepest path to the lowest elevation, and the Baton Rouge–New Orleans–Birdsfoot course is not that. Every millennium or so, the river has found a new outlet to the Gulf, abandoning a superannuated delta like this one and forming another. It is ready to move again, but the nation that de Soto explored and Jefferson defined can't afford to let that happen. Without the Mississippi, New Orleans and Baton Rouge would become salty silting backwaters, and without the river serving as trade route and freshwater supply, a large swath of the nation's agricultural and industrial base would collapse.

The Atchafalaya River is where the Mississippi wants to go. Instead of adding to the master stream, as most rivers do, the Atchafalaya takes from it. While the Mississippi heads southeast, the lesser river proceeds almost due south through a vast, undeveloped, and virtually unpopulated swamp, disgorging into the Gulf several parishes west of the Birdsfoot. The Atchafalaya route is much steeper and 173 miles shorter than the Mississippi's current course, which is why the river was very much inclined to shift sideways under Kaiser's feet.

Before the Corps intervened, the Atchafalaya's share of the Mississippi had been growing. The Red River, which drains parts of Texas, Oklahoma, Arkansas, Louisiana, and a touch of New Mexico, once entered the Mississippi a few miles north of where the Atchafalaya exits. In the 1830s and '40s, the Corps removed a 160-mile-long logjam that had, for centuries, slowed the Red's southward flow into the Mississippi. The government also removed a jam on the upper Atchafalaya. The Red pushed and the Atchafalaya pulled, and by 1945 the two rivers had become one. The Red now pours directly into the Atchafalaya, with nary a drop entering the Mississippi. This acquisitive behavior worried the Corps. In 1950, the agency decided to petrify relations between the two rivers, and it built the Old River Control Complex to manage this relationship for the rest of time. The combined flow of the Red and Atchafalaya (now hydrologically indistinct) amounted to 30 percent of the total volume of the Red-Atchafalaya and Mississippi in 1950, so the 70–30 distribution of waters became Corps orthodoxy.

Without the controls, which were completed in 1959, the Atchafalaya would have captured the Mississippi, becoming its main channel sometime before 1970. On a spring day fifty years later, the Mississippi was almost twenty feet higher than the Atchafalaya into which it poured. In the fearsome whirlpool that Kaiser beheld, the Mississippi was expressing its desire to change course. Old River Control restrained it.

In 1987, the *New Yorker* magazine published John McPhee's "Atch-afalaya," telling the story of the battle waged by Kaiser's predecessors. Some things have changed since McPhee last visited, and some haven't. McPhee began his story by walking into Old River Lock. Lockmaster Norris F. Rabalais, seeing the writer's red bandana, declared, "You are a coonass with that red handkerchief." McPhee, a New Jerseyan, knew that coonasses were Cajuns and took it for the compliment it was.

Today Russell Beauvais runs the entire Old River Control Complex. By way of autobiography, Beauvais said, "I call myself a coonass." The fifty-five-year-old grew up behind the three-story levee protecting nearby Morganza. He was ten years old when the 1973 flood tested the controls at Old River for the first time. Beauvais's father worked on the structures then and passed the stories down to his son.

The near catastrophe of 1973 takes up a good portion of McPhee's story. At the time, the Corps distributed water from the Mississippi to the Atchafalaya in only two ways, through the Overbank and Low Sill structures. During the flood, the Mississippi—scouring and ripping at the Low Sill as it pounded down into the Atchafalaya basin—began boring a hole under the 556-foot-long array of eleven gates. One of two wing walls, which guide water into the structure's mouth, col-lapsed. If the hole beneath it had broken through to the other side, the Low Sill might have become just a few more river-rounded pieces of rubble rolling toward the Gulf. And the Corps might have lost the Mississippi.

The land around Old River is about fifty feet above sea level. The hole that the river ate into the channel bed behind the Low Sill was sounded at more than eighty feet *below* sea level. The Corps rebuilt the fallen wing wall out of rock and poured more rock into the hole. Even after these and other repairs, the Low Sill is considered permanently compromised. The maximum head it can withstand—the difference in height between the two rivers—was reduced from thirty-seven feet to twenty-two.

To maintain control at Old River, in 1986 the Corps added the Aux-iliary Structure, a brutalist row of concrete cubes atop sloped concrete piers that frame six giant Tainter gates. During the 2018 flood, the Auxiliary's operation was lopsided. Several barges had crashed into one of its gates earlier that year, ripping an automobile-sized hole through the steel. Damaged, this gate had to be pulled most of the way out of the water, focusing the day's allotment of Mississippi water through a single foaming maw. Crashing into the man-made channel leading to the Atchafalaya, the water was a pure white cacophony of monstrous standing waves. The Mississippi was colliding with two rows of fifteen-foot concrete cubes, called baffle blocks, which make up the stilling basin. McPhee called the smaller stilling basin at the Low Sill "the least still place you would ever see."

"Atchafalaya" was republished in a book called *The Control of Nature*. Beyond the specific threat of the 1973 flood, McPhee makes a larger point, which every account of the Corps' forever war at Old River has restated in increasingly obvious and sensational terms. It was true forty years ago and it will be true forty years from now. Russell Beauvais put it this way: "You're really trying to fight Mother Nature. Who's gon' to win?"

During the 2018 flood, a large sand boil surfaced behind the mainline levee at Black Hawk, a short distance above Old River, where the Red and Mississippi are only three miles apart. The Corps had rapidly brought personnel, heavy equipment, and lots of sand and rock to bear, stanching the boil before the public became aware of the *what-ifs*. (Not that the Corps was hiding anything; the press releases were there in black-and-white, but the threat of losing the Mississippi was never mentioned.) Privately, geotechnical engineers at the Corps' Vicksburg District had devised several action plans for *if*: They'd drop barge-loads of rock or huge sandbags into the breach, or raise the level of the Red to dam the incoming Mississippi before it could cut a new channel. At first, most of the water from such a breach would actually flow north.

"Even if we had a massive break, we could fix that. It wouldn't be easy, but it could be done," said Kaiser. He was standing in the sun atop the Black Hawk levee with several other river commissioners. They had stopped here to survey the scene of the boil, now covered with a perfectly beveled, 72,000-pound pad of sand and limestone. "I always struggle with the concept that it's going to be 'flash done, it's over,'" he said, referring to the commonly held belief that the Corps could lose control of the Mississippi with the failure of a single structure.

It would take more than one breach to "change the maps," agreed Norma Jean Mattei, a civil engineer from Metairie, Louisiana, and the commission's only female member.

Kaiser and Mattei were still concerned with the permanent diversion of the continent's largest river, but that possibility had been studied and engineered against for three-quarters of a century. More pressing was what they didn't know: how sediment was moving (or staying put) around Old River Control.

From Black Hawk, the commissioners and their entourage drove south on the levee-top road, past cows grazing in fields dotted with bobbing pump jacks, nothing in the distance but the green wall of the levee and the wide sky. They stopped at the Low Sill office and joined Russell Beauvais in a small conference room. In the mode of a general asking his field personnel about the situation on the ground before giving an order, Kaiser asked the gathered experts and subordinates what was going on. "I was under the impression that we are not passing the required amount of sediment?" he began.

"It depends on what's 'the required amount of sediment,'" said Ty Wamsley, director of the Corps' Coastal and Hydraulics Laboratory. Wamsley explained that, in this reach of the Mississippi, more sediment was being deposited than washed away. The bed of the big river was aggrading, or rising. A higher bed meant that something called the stage-discharge relationship was changing. Stage is the height of the river; discharge is its volume. For the same volume, the Mississippi was reaching higher on the levees now than it had in 1950. To stop this, said Wamsley, "You have to divert more sediment than a thirty-seventy split." The government had consistently maintained the 70–30 distribution of water between the Atchafalaya and Mississippi Rivers—that was canon law—but it had failed to maintain that same distribution of sediment. No one knew how much sediment was going down the Atchafalaya, but it was less than 30 percent. To get to 30, Kaiser would need to divert more water.

The impact of more silt and sand going down the Atchafalaya was hard to fathom because the current regime had already filled much of the basin with sediment. Former bayous were buried under twelve feet of mud; oak, sycamore, and cottonwood forests had replaced the cypress-tupelo swamps where crawfish once scuttled. Some experts say that the Atchafalaya basin cannot handle the project design flood because, like the dams on the Missouri, the basin's water storage space is being consumed by sediment. If anything, stewards of the basin's unique ecosystem wanted less sediment, not more.

And yet, the Old River Control Complex moved far less sediment than it was supposed to. Of the five connections between the Mississippi and Atchafalaya, one—the Old River Lock—diverts barges, while the other four—the Low Sill, Overbank, and Auxiliary Structures, and the Sidney A. Murray Jr. Hydroelectric Station—divert water. The power plant was fitted into the levee in 1990 and is the newest addition to the complex. Understandably, the plant was positioned on the river to pull maximum water with minimum sediment. During low water, it took the Atchafalaya's entire allotment of Mississippi River water, while the other structures were closed completely. Annually, more than 70 percent of the water diverted at Old River went through the power plant.

The force of moving water is its stream power. A river carries lighter silts and clays suspended in its water column and pushes dunes of heavy sand and gravel along its bed. When a stream loses power, it puts sediment down. Deltas are formed this way. As the power plant siphoned water off the Mississippi, the river lost stream power and a proportional quantity of sediment settled to the bottom, adding to the aggradation below Old River.

Continuing to answer Kaiser's question, Charles Camillo, now the commission's executive director, suggested that the overall purpose of Old River Control was open to interpretation. This was a radical statement, like a cardinal saying that maybe Communion wine wasn't the blood of Christ after all. "Is it seventy-thirty? Is it to prevent the capture of the Atchafalaya? Is it to maintain the relationship the rivers enjoyed in 1950?" Camillo thought he knew the answer: The purpose of Old River Control was to maintain 1950. But he was worried that some people in the room—and throughout the Corps—assumed that the three purposes were synonymous. They were not.

"If the purpose is to keep—we're talking over hundreds of years—the Mississippi from going down the Atchafalaya," then the current distribution of water and sediment, the nominal 70–30 split, "would not do that," said Wamsley. He agreed with Camillo that the Corps' preoccupation with 70–30 might be obscuring more important things. But, as Wamsley read it, the purpose of Old River Control was to prevent capture.

As the bottom of the Mississippi rises, the river carries less water within its banks; the levees become, in effect, lower, and the pull of the Atchafalaya becomes that much greater. The original manual for operating the Morganza Floodway—thirty miles below Old River—determined that if 1.5 million cubic feet of water per second were chugging down the river, then that discharge would rise to a height of 56 feet on the floodway's river-side gates. A flood of these dimensions and growing was the official trigger point for opening the floodway. But, in 2011, when the Corps opened Morganza for the second time in history, a discharge of 1.5 million cubic feet per second stood, instead, at 59.7 feet. The river bottom had risen that much.

The power company had committed—on paper—to moving sediment in proportion to the water sucked up by the Sidney A. Murray Jr. Hydroelectric Station. However, Colonel Michael Clancy, commander of the New Orleans District, said that the Corps had never enforced the agreement. The company had received a permit to construct a pipeline to dredge sediment from the river bottom, he said, but "They never built it." The power company and the Corps had argued about it for a few years, then, "It got put on hold because of Hurricane Katrina." The Corps had rushed to stanch the breaches, both physical and reputational, caused by the 2005 hurricane and had forgotten about the pipeline, while the riverbed continued to rise.

Clancy assured Kaiser that an ambitious assessment had been authorized looking at the interplay of the Atchafalaya, Mississippi, and Red Rivers, in terms of both water and sediment.

Kaiser vowed to get the assessment funded, saying, "We got to understand what's going on here."

For Tom Holden, one of Kaiser's senior advisors, the most important question was whether they could safely pass the project design flood.

The river bottom was rising from here to Bonnet Carré, Holden said. Risk to life and property was increasing, and "You can't build the levees higher down at Baton Rouge."

An Atchafalaya-Mississippi-Red study was also begun before Katrina, and likewise forgotten. Now fifteen years had passed. "I'm a little worried," said commissioner Mattei. Regardless of the next crisis, she said, the study had to proceed. They didn't have fifteen more years to waste.

"There are changes," concluded Kaiser, "things that we don't understand."

Ever since the Corps began intervening in the Mississippi, more than a century ago, generals and politicians had made decisions and built things that, for their four years in office, for their lifetimes, largely worked out. However, the Mississippi dealt in millennia, and for each action the Corps took, the river would react—in its own place and at its own time.

The problem facing Kaiser had arguably been created by one of his predecessors, Brigadier General Harley B. Ferguson, Mississippi River Commission president from 1932 to 1939. Like much of the Corps' river policy, Ferguson's actions had their origins in the 1927 flood.

During that flood, the Mississippi crevassed a levee at Cabin Teele, on the river's west bank. Hundreds of thousands of acres in southern Arkansas and northern Louisiana went underwater. In pre-levee times, the river had naturally overflowed onto this land by way of Cypress Creek, a small distributary that was blocked off in 1921. Cypress Creek had fed a large system of slow streams—Bayou Macon, Bayou Bartholomew, Bayou Boeuf, and Bayou Tensas—which eventually drained into the Atchafalaya. As Major General Edgar Jadwin formulated his post-1927 flood control plan, he wanted very much to build a floodway mimicking Cypress Creek. That swath of Arkansas and Louisiana made a logical spillway; most of its buildings had just been destroyed, its inhabitants were scattered, and water leaving the Mississippi there would never return, so Jadwin included a Boeuf Floodway in his plan.

Area landowners vehemently opposed all floodways, but the Arkansas farmers in the path of the Boeuf Floodway were more successful than those who fought the Birds Point–New Madrid Floodway. The Boeuf was much larger, at 1.32 million acres, and the government didn't want to pay for flowage easements on all that land. Jadwin argued that since the area was subject to overflow already, the government didn't have to pay, but this didn't satisfy the public. Landowners filed suit, calling for a halt to construction. A similar case challenging Birds Point–New Madrid had been dismissed, but this one was allowed to go to trial. The government lost, and work on the nascent project stalled. The Boeuf Floodway had suddenly become very expensive.

According to Corps lore, the new chief of engineers, Major General Lytle Brown, summoned Harley B. Ferguson to his office to appoint him president of the River Commission. Ferguson had already published a plan for lowering the stage, or height, of the river by digging cutoffs, like the one Ulysses S. Grant had attempted at Vicksburg during the Civil War. Jadwin had opposed cutoffs. "The method is too uncertain and threatening to warrant adoption," he had said to Congress; the Mississippi River Commission had maintained a "no cutoffs" policy since 1884. But Jadwin's influence was waning. Ferguson described cutoffs as a new way to manage the river without turning the waters upon anyone. Brown and Ferguson sat talking and smoking pipes until Ferguson asked his superior officer, "Do you want me to write a book or fix a river?" Brown inhaled the smoke from his pipe then exhaled slowly before replying, "Fergie, you get the hell out of here and go fix that river." Brown knew Ferguson's plan and approved of it. Now, as head of the commission, Ferguson could put his theories into practice.

The idea was simple: If the river could be straightened, water would move faster to the Gulf of Mexico. No longer would it back up in those lazy bends, the long flanks of which acted like dams. Above the cutoffs, flood crest would be lower, and because the slope of the river would increase, it would scour its bed, creating a deeper channel. The Corps' missions on the Mississippi at the time were flood control and navigation, so the cutoff program seemed perfect.

A meander begins as a slight bend in a river. Gradually, the current takes material from the outside of the bend and deposits it on the inside. The bend is exaggerated as more material erodes, moving from the cut bank on the outside to the point bar on the inside. Ferguson planned to dig a channel upstream, across the narrow neck of a meander, stabilizing the banks as he went, then blow up the final bit of earth and let the river do the rest. "If you have a bathtub full of water and want to empty it—pull the plug," he explained to his staff.

Ferguson oversaw eleven cutoffs between Memphis and Old River—at Glasscock Point, Diamond Point, Tarpley Neck, Sarah Island, and Caulk Neck, among other places—transforming a notoriously apple peel–like section of the Mississippi into a virtually unimpeded straight-away. During this time, two natural cutoffs were allowed to occur, and Fergie's successor executed three more, for a total of sixteen by 1942. The length of the river was reduced by 151 miles. At Vicksburg the stage dropped seven feet, at Arkansas City twelve, at Natchez four.

Brown hoped that the cutoff program would render the costly and politically sticky Boeuf Floodway superfluous, and it did—at least at the time. Floodwater wasn't reaching as high on the levees now that the riverbed was lower. Cutoffs were cheaper and less controversial than floodways, and they didn't make enemies of the locals by demanding that they sacrifice their land. Congress approved the trade-off, and the 1941 Flood Control Act formally eliminated the Boeuf Floodway from the Mississippi River & Tributaries Project. Money already appropriated for the floodway was spent on raising the mainline levees.

Ferguson was famous even before he bent America's mightiest river to his will. He had helped raise the *Maine* from Havana Harbor, earning a reputation as a "master of whatever he undertakes." He was admired for his obstinacy and willingness to defy orthodoxy. As recently as 2016, a book about the cutoff program was subtitled "How a Bold Engineering Plan Broke with U.S. Army Corps of Engineers Policy and Saved the Mississippi Valley."

However, just a few decades after the last cutoff was completed, potamologists—*potamos* is Greek for river—began to wonder how "saved" the valley really was. In a 1977 paper, Brien R. Winkley of the Corps' Vicksburg District tried to figure out exactly what Fergie had done to the river. Stages had fallen dramatically, but what else had changed?

Winkley determined that the Mississippi in a state of nature responded to cutoffs by widening bends elsewhere, flattening its slope and redistributing its sediment to regain its lost length. Though it was constantly changing, the Mississippi below Cairo had been basically consistent before humans began tampering with it. It measured an average of one thousand miles long during the thousand-year period preceding the Corps of Engineers, and an average of twelve hundred miles long for the thousand years before that. But for the last century or so, Western civilization had imposed its will upon the river; the Mississippi was now off balance, reacting to man-made changes. Winkley and other potamologists called a balanced river, where actions and reactions were allowed to play themselves out, a river "in regime." Over time, a river in regime seeks and finds a dynamic equilibrium. Such a river would be almost impossible to live next to, build or farm by, draw water from, or navigate on, because it would jump its banks and change course. Today's Mississippi is not in regime, and probably never will be again.

According to Winkley, the Mississippi needed between thirty and eighty years to react to one cutoff. After 1942, with all sixteen cutoffs in effect virtually at the same time, the river's reaction was beyond all human capabilities of prediction. "By controlling the major rivers, most of the hazards of using floodplain lands have been eliminated; however, these rivers have reacted and are attempting to adjust to the navigation and flood control programs. It must be realized that any river is a live entity and obeys natural physical laws, and when the regime of a river is altered, there will be a response by that river," wrote Winkley. On balance, he believed that Fergie's cutoffs had made navigation and flood control more expensive and more difficult. He didn't explicitly call them a mistake, but he wrote that, judging by letters and statements from that time, "the cutoff program might have been a project motivated more by selfish interests than by sound judgment."

The cutoffs, Winkley concluded, had a more immediate and noticeable effect on the river than any single event except the New Madrid Earthquakes, which shook a vast amount of sediment loose into the Mississippi. The quakes created a wider and shallower river with more islands and sandbars. This condition would have been temporary, he wrote, but the extra sediment was still working its way downstream "when man began tinkering with the river."

Forty years after Winkley's paper, the river's gradual but massive response to Fergie's fixes had become more obvious. When Ferguson cut through all those bends, he increased the slope of the river for a portion of its length. In place of a steady grade from Cairo to the Gulf, the cutoffs created a steep section in the middle of the Mississippi, roughly between Memphis and Old River. From the moment those cuts were blasted through its meanders, the Mississippi began eroding material from the top of this steep section and depositing it below. The river was fighting to normalize its slope, to regain its regime.

Fergie had just bought time, explained Ty Wamsley. The cutoffs yielded an immediate benefit while creating a problem for another place and for a future time, "and that's what you're seeing now in the lower part of the river. It's aggrading, it's coming up." Jadwin's floodways were the right approach, Wamsley said, "but people got impatient." A floodway, though painful at the time, would have protected the valley for many centuries. Instead, he said, "The expedient solution won out."

One of Fergie's cutoffs was near Natchez, Mississippi, fifty miles above Old River. The stage there dropped four feet in the 1930s. By 2019, the benefit of Fergie's fix at Natchez had disappeared; the riverbed had risen four feet and was still rising. It may only be a matter of time before all of the stage reductions that made Fergie a hero cease to exist. A Boeuf Floodway, or some alternative means of releasing water from that stretch of the river, may once again become necessary. Ferguson's cutoffs had just kicked the can about a century down the road.

Considering the Corps' actions and the Mississippi's reactions, as well as the river's own macro behavior patterns, it's impossible to draw a direct causal line from the cutoff program to the present aggradation. The areas that are aggrading now were doing so before the cutoffs. Wamsley didn't think the cutoffs were bad. What's bad, he said, was accepting the cutoffs in exchange for the Boeuf Floodway.

Aggradation invariably coincides with degradation, or erosion. The sediment piling up around Old River had to come from somewhere. If the Mississippi was building at Natchez, it must be scouring, eating away someplace else.

Wamsley's lab learned that the river was taking material primarily from above the cutoff zone, mostly between Memphis and Helena,

Arkansas. Looking at a century of river stage data along the Lower Mississippi, scientists determined that the degradation triggered by Fergie's cutoffs had moved steadily upstream. Around the year 2000 it passed Hickman, Kentucky, across the river from Dorena, Missouri. Mysteriously, the degradation above Hickman remained minimal, while below Hickman the riverbed continued to fall.

Just downstream from Hickman Harbor, Wamsley's lab discovered a "hardpoint," a ridge on the bottom of the river made of what Wamsley called "super-super dense Pleistocene clay." The scientists tried to drill down and pull up a core sample of this remarkable clay, but it was so dense that they were only able to break off chunks. They desperately wanted to study its erodibility, to figure out how—under what circumstances—it might give way. In a flume they blasted a chunk with high-pressure water for five hours. It didn't budge. They blasted it with sand-laden water for five hours. Still nothing.

Scientists speculated that the river had created the hardpoint somehow. It was, after all, sedimentary material. They *did* know that directly downstream of the hardpoint was a plunge pool, a hole at least 100 feet deeper than the surrounding river bottom. Because the Mississippi hadn't yet found a way around or through this seemingly impenetrable "geologic control," it was eating downward, through alluvium that was tens of millions of years old. A natural river would cut new meanders, aggrading and degrading from bend to bend, moving sediment laterally until it regained its sinuous shape, but the Corps' levees and bank stabilizing structures prevented the channel from migrating. All it could do was erode.

Barges and towboats had run aground on the hardpoint during the drought of 2012, and, predictably, navigation interests had asked the Corps to remove it. In their reports, the potamologists anticipated even more navigation troubles as the riverbed below the hardpoint continued to degrade, but they begged their superiors—in the circumspect language of science—not to blast away the hardpoint. No one knew precisely what would happen if that geologic control were removed, but it wouldn't be good. The reports predicted a ten-foot lowering of the riverbed up the Mississippi to the Thebes Gap, and up the Ohio River to Olmsted Locks and Dam. Banks would cave, as would miles and miles of revetment the Corps had installed to keep the river in its channel; levees might erode and need to be moved; more groundings would occur during low water, maybe more bluff erosion and sliding. More chaos and less equilibrium.

*   *   *

Mitch Jurisich navigated not by what he saw in front of him, but by using a map in his mind. He had been plying these waters all his life, and in that time the landscape had changed almost beyond recognition. Thirteen hundred miles south of Greg Stockholm and the Lewis and Clark Lake delta, Jurisich stood at the throttle of an open fiberglass skiff. He skidded through a narrow bayou, across a lake, and into another bayou. But the shores of these water bodies existed now only in his memory—and on his radar display. Looking up from the console, he saw a vast expanse of open water.

The National Oceanic and Atmospheric Administration, responsible for surveying the nation's coast, had officially removed from its charts many of the bays and bayous Jurisich knew as a boy: Bayou Fontanelle, Bayou Long, English Bayou, Cyprien Bay, Skipjack Bay, Bay Crapaud, and Scofield Bay were gone. In their place, the charts showed only pale blue water with annotations in small print: Piles, Well, Unsurveyed, Foul, and Obstns (obstructions). Bay Adams and Bastian Bay were still there, but their outlines were shredded and porous. This was Louisiana's vanished coast, south of New Orleans and west of the Mississippi River.

Jurisich was an oysterman. Fifty-five years old, he wore sweatpants and white rubber boots. His jowls followed his neck into an oversized green T-shirt, which hid an oversized torso. He didn't dredge or shuck anymore; he ran the operation, dealing in oysters by the million, with dozens of boats working for him.

On this balmy February morning, he had the leisure to take Amos Cormier III, a forty-three-year-old lawyer and the president of Plaquemines Parish, out to the oyster beds. Plaquemines is the southernmost parish in Louisiana, where the Mississippi meets the Gulf. Most of the water bodies NOAA erased from its charts are in Plaquemines; so is most of the state's oyster industry. Dave Cvitanovich, a sixty-two-year-old retired oysterman and a friend of Jurisich, joined them in the boat. Both the Jurisich and Cvitanovich families came from the Dalmatian coast, in present-day Croatia. Throughout the nineteenth and twentieth centuries people arrived in Plaquemines from the lands that were—for a short time—called Yugoslavia. Already familiar with boats and fishing, they took up oystering and came to dominate the local industry.

Almost anywhere else, hills, islands, and fields might seem immutable, but not in Plaquemines Parish. "The landscape has changed drastically around here. We don't lose as much land no more, because we don't have the land to lose anymore," said Jurisich, surveying the water. There were many places in coastal Louisiana where marsh was disappearing by the football field, but "right here, there's not many football fields left to disappear." When Jurisich was a kid, 75 percent of this shallow, wide-open bay was land.

"Prior to Katrina," said Cvitanovich, "this area, you had a hundred houses, camps." In Louisiana, a camp is a cabin for fishing, hunting, vacationing. A community once gathered here each summer, miles by boat from the nearest town, to harvest oysters and cook Croatian meals. They had enough dry land for kitchen gardens. Only one camp remained. On stilts.

"I know exactly in my mind what it looks like. I know the old bayous, the old channels, and we still have them marked with these poles, just like they used to exist," said Jurisich. As he sped along, thickets of poles—bamboo and plastic—whizzed by on either side of the skiff. The poles staked out legally binding oyster leases—rectangles of water bottom, rented from the state and passed down through generations. Some were more than a century old. The oyster farmers painted the ends of their poles with an identifying color, the way lobstermen paint their buoys in Maine. Mitch Jurisich's had yellow tips.

Cvitanovich described how, when Jurisich was a child, he and his family had weathered Hurricane Camille nearby, floating on the bayou. Around here, hurricanes mark time. Storms are recalled with the familiarity of ex-girlfriends—if everyone had the same ex-girlfriends: before Betsy, after Camille, during Rita, after Katrina. (Old boyfriends are less important: Isaac, Ike, Gustav.) Jurisich's father had decided that the family would be safest in their boat during Camille, tied to the pilings in front of their camp. It was Mitchie's sixth birthday. He played with his new Hot Wheels cars in the boat's cabin while the wind howled outside. The storm would kill hundreds of people and cause billions of dollars in damage.

"You weren't scared?" asked Cormier.

"Nah," Jurisich said. "I didn't know no better."

All of Louisiana's buildings and levees, even its marshes and the Mississippi River itself, sit on top of a miles-thick wedge of ancient river-borne sediment. Some iteration of the Mississippi built it all, from Lester Goodin's soybean fields to Mitch Jurisich's oyster beds. The wedge rests on the edge of the continental shelf, which is slowly sinking under its immense weight. As water is squeezed from between grains of sand and bits of plant matter, the accumulated mass compacts. This sinking and compacting is called subsidence.

The river and its tributaries have always moved sediment downstream. For about seven thousand years, the constant depositing of new sediment built enough land to offset the effects of subsidence. This began to change in 1717, when the first levee was built in front of New Orleans. The French wanted to keep the sediment-laden waters off their land. Eventually, levees lined both banks of the Mississippi, from the Benton and Kentucky Hills, all the way to the Gulf of Mexico. The river was well leveed before the Great Flood of 1927, and thoroughly straightjacketed after.

In the 1800s, the combined Mississippi-Atchafalaya River system moved about 400 million tons of sediment to coastal Louisiana every year. In the middle of the following century, the Corps dammed the Missouri and Arkansas Rivers, which once brought the dusts and sands of the Rockies and Great Plains to the Gulf. These days, the Mississippi-Atchafalaya system moves around 150 million tons of sand and silt per year, a 60 percent decrease.

Most subsidence is natural; most dams and levees are not. These civilizing constructions fostered agriculture, industry, and commerce, but they also curtailed the supply of sediment that once built up the

Gulf Coast faster than it could sink. A lot of Louisiana is already gone. The rate of land loss accelerated from forty-three hundred acres a year in the early twentieth century, to almost ten thousand in the late 1940s. It peaked during the 1970s, when almost twenty-five thousand acres a year were lost. The number now hovers around ten thousand. Miles of marshland and dozens of islands disappeared; roads and houses and whole towns sank. Losses were measured not just in football fields, but in Manhattans and Delawares.

Louisiana began to pass environmental restoration laws in the 1970s and began some modest restoration projects around 1990. Then came Hurricane Katrina. Three weeks later, an even more powerful storm, Hurricane Rita, hit Louisiana's less-populous southwest coast. To defend against similar disasters in the future, the State of Louisiana embraced a radical solution: Let the river out. Reversing a century of struggling to contain the Mississippi, the state now wanted to mimic Mother Nature by diverting the Mississippi and reconnecting it to its natural delta. As the river spilled out of its levees and slowed down, it would drop sediment and build land. Martinsburg shale from Appalachia and Pikes Peak granite from Colorado would once again contribute to the mass of Louisiana.

By diverting the Mississippi in a controlled way—partially recreating its premodern condition—Louisiana's Coastal Protection and Restoration Authority hoped to rebuild the lands that had once served as buffers between population centers and the winds and storm surges of the Gulf. The most ambitious of these projects was the $1.4 billion Mid-Barataria Sediment Diversion. Designed to reconstruct the marshes south of New Orleans, it would divert up to seventy-five thousand cubic feet per second of freshwater into Barataria Bay. A concrete canal would cut through the Mississippi River levee to give the river an outlet into the marsh. A gate would control how much water was diverted and when. At peak flow, the diversion would rank among the twenty largest rivers in the United States.

Few states could afford to spend billions of dollars to restore ecosystems without federal help, least of all a relatively poor state like Louisiana. Ironically, the two worst things that ever happened to Louisiana's coast—Hurricane Katrina and the BP oil spill—propelled billions of dollars into state coffers, giving the state an unprecedented opportunity to rebuild its coast and do it fast.

A river diversion like Mid-Barataria had never been attempted. Debate centered on how much land could be built and how quickly, but no one doubted that huge infusions of freshwater would change ecosystems. The diverted Mississippi River would flow out through Barataria Bay toward Bay Adams and Bastian Bay, home to some of the state's best oyster beds. Oysters need brackish water—not too salty, not too fresh; if they're exposed to overly fresh or overly salty water for extended periods, they die—and they can't move when their environment becomes inhospitable. Barataria Bay also sheltered a pod of dolphins. If the water became too fresh, scientists determined, the dolphins might not survive. Congress had given Louisiana a one-off exemption from the Marine Mammal Protection Act, a special dispensation to kill dolphins. The state seemed prepared to sacrifice the dolphins and its oyster beds to protect metropolitan New Orleans.

Mitch Jurisich gently pressed his skiff against the side of one of his oyster dredges. He stayed on the skiff as Dave Cvitanovich stepped aboard the dredge with Amos Cormier, the lawyer and parish president. Jurisich's son Nathan was standing behind a console at the stern of the square, open-decked aluminum boat. On his left and right were two metal tables, each with two men in rubber overalls, boots, and gloves. The dredge swerved over the beds in lazy curves, dragging two steel rakes. Nathan pressed a button. Winches turned, chains clanked, and both rakes broke the surface. Oysters poured onto the tables in a roaring clatter that became a whacking staccato as the men broke up the muddy lumps with little hatchets. They knocked old shells and little oysters onto the deck, and tossed oysters three inches long and longer into a bucket. The rakes dropped back into the water in a blur of slithering chain. As the buckets filled, the oysters were dumped into coffee sacks—"Coffein Compagnie" and "Bremen, decaf"—and dragged to the front of the dredge, which was already a third full. Budweiser cans and Marlboro packs mixed with the shell litter. The dredge's canvas top flapped in an offshore breeze that threatened rain.

Cvitanovich grabbed a handful of fresh oysters. With a calloused thumb he wiped away the mud to reveal a spat, a baby oyster, the size and consistency of a fingernail. Nicely shaped, symmetrical oysters were for the half-shell trade, to be eaten raw; the weirdly shaped ones were for frying. Using a shucking knife, Cvitanovich deftly opened an ideal oyster. It tasted clean, fresh, and barely briny. These oysters were skinny, he said, because the river was rising, and freshwater made oysters shrink. Though there weren't any diversions in this area, it was close enough to the Mississippi's mouth to be influenced by the flooding river, which, running along between its levees, was easily six feet higher than the surrounding marsh.

On a good day, Jurisich could see forty-five boats out on his leases, and fifty families working for him. Most were sharecroppers who harvested oysters, sold them, and split the proceeds with him, fifty-fifty. Sunburned Alabamans on narrow boats scooped up oysters with long tongs. People in wetsuits walked over the beds "cooning" for oysters— gathering them up by hand, like a raccoon. Tongers gleaned leases that had already been dredged. Cooners picked them over again. And then the oyster farmers withdrew from the lease; it would take two years for a new crop to reach market size.

Jurisich's workers would land two thousand sacks by the end of this February day: 360,000 oysters. He frequently harvested a million oysters a week. The day's dockside price for half-shell oysters was $60 per bushel; $44 for a frying sack. Jurisich stood to gross at least $88,000 before the sharecroppers got their cut. Cvitanovich reckoned that his friend was among the top five oyster producers on the Gulf Coast. "Mitch's got about a seven-million-dollar yacht, got a crew on it and everything else," he said to Cormier later, "but did you notice that today? He works for what he has."

Cvitanovich and Cormier got back on the skiff, and Jurisich throttled forward, navigating by memory again, with a glance at the radar screen. "All inside of this line of poles, that all used to be land: Bay Cheri, Bayou Cheri, from there we get into Bayou La Chute," Jurisich said. He slowed beside a cluster of blackened pilings. Just breaking the surface was a mound of shucked oyster shells with a seagull standing on it: a midden left by the Jurisich family. Mitch's grandfather had come over in 1904, and built a house on this spot in 1920. This had been their camp.

"They sustained life here," said Cvitanovich. "They had—seriously— fig trees. They had pomegranates. They had chickens. They had goats." He could remember acres of solid, dry land here, four feet above the highest tide line. Now it was all four feet below.

As kids, Jurisich and his cousins canoed around the bayous in Cajun pirogues. The land was sinking fast then, and they noticed small oysters living on the newly submerged marsh. By law, underwater land became state property, and the Jurisiches were losing property by the acre. In 1976, Jurisich's father asked the state to lease him a stretch of the new water bottom. The state told him, "You must be crazy but, yeah, we'll lease it to you," his son remembered. Wherever they worked, oyster farmers seeded the ground, waited for their oysters to mature, harvested them, and seeded again. They weren't depleting any resources—oysters remove contaminants from water—just harvesting an annual crop. In those days, oysters were farmed in the deep channels. Shallow-water farming was something new. "We evolved with the times," said Jurisich. The idea caught on, and the state put thousands of acres of sunken marsh-land up for lease. At first, a hundred acres of flooded marsh yielded as much as one acre in the bayou. But the industry adjusted its methods, and now the yields almost matched.

"The state never stepped up, never wanted to do nothing," Jurisich said, looking out at his family's mound of shells. "Coastal restoration wasn't cool back then. So no matter how many times we complained about what was happening, it fell on deaf ears. There's billions and bil-lions of dollars' worth of jobs available in that field now, and what all these billions have created is a bunch of NGOs and kids getting out of school thinking they have a solution to the problem.

"So what the state would love to do is come in and fill all of this up, flood us with freshwater hoping to build new land. Problem is, it doesn't give us evolution time. If they start flooding us, we're going to lose our oysters. Now all these new people come in and they got all these bright ideas? Well, hold up a second. We built what we have with these hands—and with no help."

"That set of pilings means a lot to us," Jurisich said. He turned the wheel of the skiff and throttled down, slicing off across the lost land.

Several diversions already exist along the unpopulated, unleveed riverbank near the Birdsfoot, though none were constructed with land creation in mind. During the 1973 flood, the Mississippi cut through its natural bank and began gushing into the marsh near Fort St. Philip. Farther north, Mardi Gras Pass formed during the flood of 2011. The pass was christened by John Lopez, director of the coast and community program for the Lake Pontchartrain Basin Foundation. Mostly thanks to Lopez, Mardi Gras Pass is the best-documented diversion in existence, touted by the state and environmental groups as proof that the unfettered river can create land.

In the 1960s and '70s, saltwater was intruding so far into the marsh that the oyster farmers actually asked the government to dig canals with gates at their mouths to draw freshwater from the Mississippi into the marsh. One of these was the Bohemia Salinity Control Structure. In 2011, the flooding river cut around the control structure and drastically widened its channel. Lopez was there, standing in thigh-deep water as the pass formed. He watched as it eroded upstream, "like a vertical ledge eating away toward the river." The pass was controversial enough that Lopez said, only half joking, "We never used a shovel"; the cut was entirely natural. In what the oyster industry sees as a deliberate act of rebranding, Lopez named the new, uncontrolled diversion Mardi Gras Pass, after the holiday during which it was born.

On a blustery February day two weeks after Mardi Gras in 2018, Lopez stood at the helm of a fiberglass motorboat churning toward the pass. Approaching its mouth, the boat rounded a knob of land that bulged fifty feet into the adjacent canal. Nosing into the mud at the edge of the knob, Lopez cut the engine. Posts that had once marked the entrance to the pass were now several yards inland and barely visible, overwhelmed by shaggy vegetation. The ground was firm; a ledge of black soil led to tall grass and small willows. This whole knob, over an acre, had appeared since 2012. It was growing outward by a centimeter every month, five vertical feet per year. "There are people who flat out

say Mardi Gras Pass hasn't built land and you can't stand on it," Lopez
said, defiantly. ("Get John Lopez and go walking next week on the land
that's supposedly built," one oyster wholesaler had said. "You'll go up
to your neck in muck.")

Lopez was sixty-three years old, short and burly, with gray-streaked
hair and a thick beard. A native New Orleanian, he had worked for
twenty years as a geologist in the oil and gas industry before quitting
to pursue a doctorate. Then he joined the coastal-restoration branch of
the U.S. Army Corps of Engineers. A "closet environmentalist" all the
while, he also volunteered with the Lake Pontchartrain Basin Foundation. Two months before Katrina, the foundation hired him to run its
coastal sustainability program.

Lopez lives in Slidell, in St. Tammany Parish, on the north shore of
Lake Pontchartrain, a tidal estuary cut off from the Gulf by an ancient
river delta and still open at its eastern end. Driving through St. Bernard
Parish on his way to Mardi Gras Pass, he passed abandoned houses flat
on the ground, and new houses on twenty-foot stilts. After Katrina, the
federal government paid to raise flood-prone houses. Older ones could be
jacked up for about $30,000—"a weekend project," said Lopez. Newer
ones, built on concrete slabs, cost three times as much. Lopez knew this
because Katrina had destroyed his house, too.

"We came back, the house was gone, just washed away," he said. The
property he and his wife owned stood three feet above sea level. Katrina's
storm surge was twelve feet, "and that doesn't count the waves." Lopez
and his wife moved inland, to a small town near Baton Rouge, while
they rebuilt. It took four years. "I don't think we made a rash decision
to move back," he said. "The place we felt like we wanted to be is
where we were." The new house, on the same site, stands twenty-two
feet above sea level, on concrete columns. They park their cars on the
slab beneath the house, and take an elevator to the first floor. During
Hurricane Isaac in 2012, only the base of the columns got wet. Lopez
and his wife evacuated, to be on the safe side, and returned home in
four days.

CPRA's Master Plan had redefined the region's habitable land by
publicly identifying which areas it could protect and which it could
not. Nearly three dozen tracts remained beyond the shelter of new

construction. These areas, outside the levees, were deemed too risky to live in, even when houses rose up on stilts. (A house raised too high, Lopez pointed out, could be blown over by the wind.) The state had proposed "non-structural" protection measures for these areas. The house that the Lopezes rebuilt stands in one of these non-structural zones. The state wouldn't build levees around them, nor would it force anyone to move. It did hint that a few thousand houses might be bought by "voluntary acquisition," but it couldn't say exactly where those houses might be. As Lopez put it, "They're not saying you've got to relocate. They're not saying you have to abandon your land. But basically you're being put in that category."

Lopez wasn't bitter; he'd been an architect of the policy that now left him in limbo. CPRA based its Master Plan on a relatively new strategy calling for multiple lines of defense against a storm's onslaught. The new plan moved further from the "levees-only" past than ever before. The lines of defense are like a series of speed bumps, impeding wind and water: Barrier islands, salt marsh, natural ridges, fresh marsh, solid land; only far inland would earthen levees and concrete walls rise up as a last line of defense.

To create habitable land out of swamp, the early developers and occupants of New Orleans built levees to keep the river out, then drained the encircled land. The Dutch call these bowls with no natural drainage "polders." Rainwater collecting inside ring levees has to be pumped out. Because the floors of these polders consist of fine, peaty soils—not rock or clay or anything hard—pumping makes the land sink further. Years of pumping helps explain why some areas of New Orleans are eleven feet below sea level. The Mississippi in flood can be twenty-eight feet above homes in these neighborhoods.

More than half the population of New Orleans evacuated during and after Katrina. At the time, some informed voices asked if the city's remaining residents might be wise to move—permanently. A planned retreat from the city seemed possible. On a long enough timeline, wasn't New Orleans doomed? Maybe, but no politicians—nationally or locally—would say so. "We will stay as long as it takes to help citizens rebuild," said President George W. Bush two weeks after the storm. "This great city will rise again." Though retreat might have been the prudent and rational thing to do, "It's just wrong," said one oysterman. "This is the United States of America. You don't allow that to happen to one of your cities when you have a major disaster." Americans have never enjoyed admitting defeat, especially when the adversary is Mother Nature. A planned retreat would also have required complex expropriation procedures and billions of dollars for buyouts, neither of which was forthcoming. Politicians chose the path of least resistance: Let everyone rebuild, and wall off the town. Besides, the areas most affected by Katrina, the prime candidates for abandonment, were home to much of the city's Black population; it was the Pinhook conundrum magnified.

More than one million people now live in the New Orleans metropolitan area, and the state and federal governments are scrambling to protect them. Louisiana has reconstructed most of the barrier islands that once lay in the Gulf south of the city—the first line of defense—by piling up the river's own sediment. They've rebuilt some marshland by pumping sand and mud into enclosed areas, but the process is expensive, and to offset subsidence and erosion, it will have to be repeated ad infinitum. The only way to beat these forces of degeneration, the state believes, is to build with the sediment-laden Mississippi.

Though Amos Cormier was the parish president, he was the junior man on Mitch Jurisich's skiff. He listened respectfully to the oyster farmers, and asked them more than once what he should push for in a scheduled meeting with Louisiana's governor. As Jurisich and Cvitanovich had hoped, Cormier seemed to absorb the idea that seafood was (borrowing a buzzword from the NGOs) the only "sustainable" industry left in his parish. By one estimate, Louisiana harvests 850 million pounds of seafood each year—second only to Alaska—with an economic impact of $2.4 billion annually. How big a bite would the diversion take?

Cormier might have fit the part of a fast-talking litigator—well dressed, graying at the temples, sharp, acerbic—but in Plaquemines he benefited from a deep well of credibility. His grandfather was an Acadian who arrived in the parish on an oil and gas company's seismograph boat. His grandmother was French-Croatian, from a seafood family that loaded their house onto a barge—after Burrwood, where they used to live, began to sink beneath the Gulf—and floated it to Buras. Fuel and sulphur took over the parish early in the twentieth century. Freeport Sulphur once had a thousand employees, mostly in and around Port Sulphur, where Cormier grew up. Oil and gas strikes in the marsh triggered heavy investment in pipelines, canals, and wellheads. An era of massive production followed, creating a lot of jobs and a lot of environmental damage. By the time Betsy hit, in 1964, the sulphur was beginning to play out and the inshore oil reserves were running low. The industry moved out to sea and consolidated its base of operation farther west, in Lafourche Parish. Just a few oil-related helipads and docks remain in Plaquemines now, near Venice, the last town you can drive to on the Mississippi.

The name "Plaquemines" is derived from an indigenous word for persimmon. Orange groves once lined the riverside, and the title of Orange King, bestowed during the Orange Festival, is still a big deal. (Jurisich was named king in 2014.) Throughout the booms and busts, there was always seafood: oysters, crabs, and lots of fish. Cormier's grandmother told him that even during the food shortages of the Great Depression, people in Plaquemines ate well. His father, a schoolteacher and beloved coach, served as the parish president until 2016, when he died in office at the age of seventy. Cormier won a special election to replace him. Both father and son ran as Republicans.

Over oyster po'boys at the Black Velvet Lounge in Buras, Cormier spoke about the Mid-Barataria Diversion and what it would do to the parish. "My biggest fear with this," he said, is that "they've got no empirical data to say that this thing'll work, and if you look historically, we always get screwed. Plaquemines always gets screwed." During the 1927 flood, bankers and power brokers in New Orleans, with state and federal approval, decided to relieve pressure on their city by dynamiting the levee at Caernarvon, in neighboring St. Bernard Parish. To stop them, people in Plaquemines and St. Bernard began patrolling the levee with loaded guns. They backed down only after receiving a promise from the New Orleans bankers that Cormier characterized as "Don't worry, we're going to compensate you." The levee was dynamited, inundating St. Bernard and the east side of Plaquemines, but the promised compensation never came. The Mid-Barataria Diversion, he said, "reeks of the same scenario." CPRA was offering the same kind of vague promises. Cormier wasn't aware of any money set aside in escrow for mitigation. "We're just supposed to believe 'Trust me.' No, I'm not going to trust you, because I know the history of what happened before."

In premodern times, when the Mississippi River flooded, the heaviest sediments dropped out as soon as the river overflowed its banks, forming natural levees. The highest ground in the delta is actually right beside the river, its ancient courses, and former distributaries. The French founded New Orleans on the Mississippi's natural levee. Almost all the settlements in south Louisiana stand atop these slight ridges. From there, the land slopes downward. When the French apportioned property, they divided it into narrow slices, extending from river to marsh. They measured it in arpents, about two-thirds of an acre. Most properties were three or more arpents wide and forty arpents deep. Houses were built on the high ground facing the river, which was the main thoroughfare until after World War II. Farms occupied the sloping land behind the natural levee, and trappers worked the wetlands behind that. About forty arpents from the river, the ground became very soggy. At that point, the French dug canals to help the land drain. After Betsy, back levees were built on the marsh side, around the twenty-arpent line. A contemporary map of Plaquemines Parish, minus the marsh, would amount to two strings of land flanking the Mississippi in its final run to the Gulf. The federal government in the form of the Corps oversees the river side; levee districts—now under the umbrella of CPRA—are in charge of the marsh side. The feds and the state have a history of imposing their will on the parish, and levees on both sides have been bulldozed and rebuilt, narrowing the strings of land even further.

Cormier felt squeezed. "The Corps keeps moving the levee back," he said. He'd finished his po'boy and was sipping an Arnold Palmer. "After Katrina, they came down here and they were moving it back again. My dad told the Corps, he said, 'What the hell are y'all doin?' 'Oh, we're doing this for the good of the parish, we're moving it farther back.' My dad said, 'Y'all been taking land ever since I was a kid over here. Why don't y'all move it back in the French Quarter over there?'"

Jurisich couldn't come to lunch, but Cvitanovich had devoured a tuna steak and was enjoying a glass of wine. "The front steps of my daddy's barroom is where the toe of the levee is now," he said. "Just the parking lot of my daddy's barroom is still there. Where the barroom was is in the river." Cvitanovich described a second property he owned. In the title, it specified "six hundred and seventy-five feet from the toe of the road to the levee. Know how much I got right there right now? Sixty feet, from the toe of the road to the base of the levee. That's how much the levee has moved in." He said he could remember when the Mississippi was 250 feet wide; he knew a man who claimed he could throw an oyster shell across it. Now that man would have to throw his shell half a mile.

Shipping interests, the Corps, the port, and the city of New Orleans all benefited from a wider river. Plaquemines suffered. In Cormier's view, the Mid-Barataria Diversion was yet another threat. He'd done everything he could, in reams of lawyerly letters, to thwart its construction. Exercising what limited powers he possessed, he forbade the state's scientists from taking soil borings within the diversion's footprint. CPRA threatened to sue. When Cormier didn't back down, CPRA claimed to have found a loophole that allowed it to take the borings without parish consent. "Mao Tse-Tung would be proud," said Cormier. "They did it with the barrel of a gun." Armed state troopers escorted the borers, but no confrontation broke out. The borers bored.

The site of the Mid-Barataria Sediment Diversion is just north of the historically Black town of Ironton, a four-street grid of neat, prefab houses and one church, sheltering behind the levee on the Mississippi's west bank. North of the town, unruly trees crowd the levee on both sides. This anonymous swath of woods was once part of the massive Myrtle Grove sugar plantation. Now it will become the diversion. Two miles upriver from the site is the Phillips 66 Alliance Refinery—a forest of white tanks and fuming chimneys surrounded by a private levee and a barbed wire fence. To its west lies cow pasture and, beyond that, marshland.

Both John Lopez and Mitch Jurisich blamed oil and gas for some of the delta's sinking. Subsidence had accelerated in the 1970s, when inshore extraction reached its peak. Companies pumped millions of gallons out from beneath the squishy mass that passes for land here. To reach their wellheads, they also cut canals every which way across the marsh, chopping up its surface while sucking out its insides. Wave action, not to mention hurricanes, eroded the canals. Some grew to ten times their original width. Others, merging with the fraying edges of natural lakes and bayous, melted away altogether. Grass-tufted spoil banks, mounds of dredged marsh bottom, are often all that remain above water. To the oil and gas companies, the seventies subsidence bump was just Louisiana sinking as usual. However, geologists couldn't see how a natural phenomenon would spontaneously speed up and then slow down again. They couldn't prove causation, but the correlation was clear.

On the east side of the river, in the emerging wetlands being created by Mardi Gras Pass, a hiss was coming from a clump of pilings and a tangle of rusty pipes. There was no smell. Lopez had reported this gas leak months ago. The Coast Guard told him it had been fixed. It hadn't. Lopez was out in his motorboat again, checking on the progress of Mardi Gras Pass. He'd visited in February. Now it was November.

He turned his boat around and took a left, down another oil and gas canal. The landscape was bright green. While most of these old canals were growing wider, this one had narrowed; Mardi Gras Pass was at work. Mudflats on both sides of the canal broke the water's surface in matte streaks. Pelicans and gulls stood on the flats. Thick tufts of grass grew behind them. A flock of bright pink roseate spoonbills took flight as the boat slowed. Snakes swam by.

Contrast this with the lone trunks of dead oak trees, their roots killed by intruding saltwater, which had become icons of the vanished coast. Or stands of dead cypress, branches intact but white and brittle.

Lopez described Mardi Gras Pass as the main stem of a new delta. It was widening and deepening the old salinity control canal, as well as pulling sediment from the Mississippi, which it carried into various branch canals and channels. Unlike the Mississippi, which "wastes" much of its sediment, sending it off the continental shelf, the pass captured all that it took and distributed it to lakes and bays dozens of miles away.

In 2017, Mardi Gras Pass was swallowing 1 percent of the Mississippi's full flow; by 2018 that had doubled. Could it ever divert the entire river? The Corps seemed unconcerned. For the time being, it was content to watch it progress.

Bren Haase, CPRA's head planner and an architect of its 2017 Coastal Master Plan, could refute all the arguments against diversions. He could convince you that rocks and dredging wouldn't go far enough; that the river wouldn't run out of sediment; that salt marsh wasn't better than fresh. But for Haase, such arguments were beside the point. "It's no longer a debate about this or that," he said. "We gotta do all this stuff." Louisiana urgently needed to protect itself by every means possible. Katrina was not a one-time event. "Nobody who lives along our coast, thinks it's not going to happen again," he said.

The land loss forecast in an earlier Master Plan, published in 2012, was "kind of grim," he said, but, to his surprise, the public hadn't reacted in anger. The plan only formalized what everyone already knew, Haase said, which is that "we are in a state of emergency."

With or without diversions, change would come to the coast, as it always had. "Whether we implement the projects in the Master Plan or not, the future of our coast is going to be extremely different, and that's going to be really good for some resources and really bad for some resources," he said. Salinity would change, vegetation would change, the sea would rise, and the land would sink.

John Lopez liked to show a photo of oysters growing on the knees, the knobby protruding roots of cypress trees. Cypress are freshwater trees; salt kills them. The shells on the knees meant that as the water became salty, the trees died and the oysters took root. If the water became fresh again—via a diversion, perhaps—the oysters would die, but the loss would be part of a cycle. The landscape has never been static. Lopez admitted that the diversion would kill some oysters. "In Barataria, yes, it will shrink the area of productivity," he said. "But will it mean fewer oysters? I'm not convinced." Eventually, new beds might be established at the new fresh-salt water frontier. Besides, what was the alternative, Lopez wondered. "Manage for oysters while the whole place falls apart?"

The productive grounds on the west side of the river, where Mitch Jurisich had his leases, were usefully salty. Jurisich owed his success to an altered and unstable ecosystem, as he well knew. The Master Plan's projections for land loss and sea level rise showed that the water in

Barataria would eventually become too salty for oysters if no action was taken. Haase felt that the oyster farmers and others blamed CPRA for trying to change Barataria Bay. Yes, CPRA sought to change the bay—for the better, he would argue—but the choice was not between change and no change; it was between two different changes, and neither outcome could be predicted with particular accuracy.

To study something as complex as the Mid-Barataria Diversion, you need to build a physical model. Computers have a hard time with chaos, and when fast-moving, sediment-laden water hits a fixed object, hydrological chaos ensues. Even the most powerful computers can model only a few seconds of such a scenario. There are just too many variables. To figure out how (and if) the diversion would work, CPRA paid Alden Research Laboratory $4 million to build a one-sixty-fifth scale model of the diversion inside a warehouse-sized laboratory in Holden, Massachusetts. To replicate the sand that the diversion is supposed to deliver to Barataria Bay, the researchers at Alden designed a custom particle, mimicking the weight and feel of the real thing: "white acrylic with the consistency of sugar, ordered in vast amounts and costing about a thousand dollars a ton," wrote John Schwartz in the *New York Times*.

Ty Wamsley modeled the complex interaction between structures, water, and sediment in several hangar-size Quonset huts at the Corps' Coastal and Hydraulics Laboratory in Vicksburg. The Mississippi's bottom is among the most dynamic environments on earth. Science understands more about the Mariana Trench—the deepest part of the world's oceans, at thirty-six thousand feet—than it does about the two-hundred-foot-deep holes in the bed of the Mississippi. The river's extreme sediment loads, dangerously fast currents, and zero visibility have made it impossible for anyone—not even a robot—to plumb its depths. Scientists don't even know why these holes in the riverbed exist; computer models say they should fill with sand.

On a warm fall day, Wamsley got in his car for a drive around the laboratory's grounds. Built soon after the flood of 1927 as the Waterways Experiment Station, this was the first scientific facility devoted to understanding the river. In the 1940s, German prisoners of war were put to work building a two-hundred-acre model of the entire Lower

Mississippi, from Cairo to the Gulf. The lab used the model to design and test the Old River Control Complex, before the rise of computers prompted them to abandon the physical replica. But because fluid dynamics can overwhelm even the best computers, Wamsley's lab still builds one-off models of specific structures and environments.

Wamsley stopped in front of a double-hulled aluminum boat up on a trailer. A hinged arm between the hulls held a long steel pipe. More sections of pipe lay on the ground. This was the Corps of Engineers' coring boat. On the river or in the marsh, the boat's coring rig drove down pipe, section by section, and brought up samples of material. The samples couldn't be moved far, since any jiggling might rearrange the particles, so a mobile flume was floated out to analyze them on-site. Removed from its pipe, the tube of mud was placed in the flume and hit with water at different speeds to analyze its erodibility. Wamsley hoped CPRA would do this kind of study on Barataria Bay. Erodibility would be critical for determining how much land would be built—or lost—when the diversion was open.

A colleague of Wamsley's had said that "as an engineer, you don't ever think that you can't do something," but you might just build yourself a new problem. "By the time you get done, it might be cheaper to just buy bags of dirt from Home Depot and throw them down there." The oyster farmers and their allies insist that diversions build land slowly—if at all. It will take several generations for a diversion to accomplish anything, they say, while the fishery will be ruined in a matter of years.

"I think they decided that they're going to build a diversion," Wamsley said. The people of Plaquemines Parish have a legitimate point, he said. "They're going to lose their living for something that might not work." As Wamsley saw it, the outcome of Mid-Barataria ranged from "moderately beneficial" to actually accelerating land loss. None of the existing diversions—not Mardi Gras Pass, Fort St. Philip, or the Bonnet Carré Spillway—were adequate analogies, he said. Mid-Barataria would build land, but it would scour land away, too; it would kill off the salt-water ecosystem; it would raise water levels. All told, would there be a net land gain? Wamsley wasn't convinced. There was also the question of how much this land would cost. It might be very expensive, maybe as expensive as pumping and dredging.

If Wamsley's scientists had a role to play, he said, it would simply be to "make sure everybody understands the uncertainty. There's a lot of dogmatism on this, that's my problem. A lot of work indicates that no one should be so dogmatic. People that are for it are so adamantly *for it*, that I don't think they understand how uncertain it is."

A sediment diversion into Barataria Bay was originally considered as part of a federal coastal restoration project. But with all the BP money, Louisiana decided to build this first-of-its-kind project alone. The state still needed the Corps, though, to approve alterations to the MR&T levees and the navigation channel, and to make sure it complied with federal laws like the Clean Water Act. An environmental impact statement compiled by the Corps would ultimately determine whether the diversion was "in the public interest." But the Corps would not tell Louisiana whether or how to compensate groups like the oyster farmers, who might find themselves on the losing end of the project. Nor would the Corps determine whether its benefits outweighed its costs.

"It's not our money," Wamsley said. "Whether or not it works is not really a judgment we have to worry about. If it's a federal project, we have to justify the investment." The entire environmental permitting process had been designed to preserve and rebuild landscapes altered by traditional construction. It was never intended to evaluate a deliberate environmental change. "When you do that, there are winners and losers," he said. "What's good for a bass isn't good for a speckled trout. Who makes that decision? It's very subjective."

Alex Kolker, an associate professor at the Louisiana Universities Marine Consortium who contributed to the 2017 Master Plan, was confident that the Mid-Barataria Diversion would build dozens of square miles of land in less than a century. A man-made diversion of the Atchafalaya, called the Wax Lake Outlet, had built about sixty square miles since it was completed in 1942 (largely under Harley B. Ferguson's supervision). Mid-Barataria would work similarly, Kolker said, depositing between a half-inch and one inch per year. In geologic time, this growth rate was superfast, but it would be "some time," he said, before there was enough land to walk on, before the lines of defense were reestablished and New Orleans was better protected.

The diversion could be operated to maximize land building while minimizing freshwater input, Kolker explained. If it were opened only in late winter and early spring, it could capture the high sediment loads of the flooding Mississippi with minimal damage to the oyster reefs, because cold water causes less harm to the bivalves than warm water. The best time to close the diversion would be in late summer, when the river was warm and sediment-poor. The structure could also be opened selectively to grab the sediment peak, a period of a few weeks that seemed to follow just behind the flood crest, then closed once the peak had passed.

Another question was how diversions would affect the river's stream power, a finite resource, like its sediment. Louisiana's Master Plan ultimately called for four big diversions below New Orleans. Say each diversion took sixty thousand cubic feet of water per second away from the river, which might have a total volume of 1 million; they would be consuming 24 percent of the river. How much more sediment would drop out in the channel when that stream power was lost? Already the Corps dredged almost constantly at Head of Passes, where the Mississippi branches off into its Birdsfoot, mostly to keep the crossing deep enough for oceangoing ships. If the diversions were run only during high water, there would be plenty of stream power to go around. During low water, when there wouldn't be much sediment anyway, diversions could be in conflict with navigation over stream power, and hypoxic zones—areas of low-oxygen water, harmful to plant and animal life—could become a problem.

Sea level rise was also a huge variable. A higher sea would decrease the slope of the river as it moved past New Orleans and out to the Gulf. A flatter river would likewise have less stream power and would deposit sediment in different places. So far, subsidence is measured in feet, while sea level rise is measured in inches—but the sea is rising rapidly, while subsidence is gradually slowing. Subsidence brought coastal Louisiana to where it is today, but sea level rise will be responsible for its future.

As a river abandons one delta and chooses another, the old delta naturally transitions from a freshwater regime to a saltwater regime; the old delta sinks and is inundated while the new one grows. The mouth of the Atchafalaya is the freshwater-dominated beginning of the Mississippi's next delta, while the delta of Plaquemines Parish is sinking and becoming salty. Engineers may be able to keep the Mississippi from changing course, but the river's current delta is dying, and nothing can stop that.

Scientists at Tulane University concluded in a 2020 paper that the collapse of Louisiana's coast was probably irreversible. The accretion of coastal marshland was not keeping pace with relative sea level rise. Looking at the growth and submergence of deltaic plains over the last 8,500 years, the researchers determined that the remaining Louisiana marshland could convert to open water in as little as fifty years. Torbjörn Törnqvist, the paper's lead author, speculated that the new shoreline would be around the latitude of Baton Rouge and the north shore of Lake Pontchartrain. Everything south of that—New Orleans and most of its suburbs, all of Terrebonne, Lafourche, St. Bernard, and Plaquemines Parish—would be gone. "We're screwed," Törnqvist said, paraphrasing his findings in an interview with Mark Schleifstein of the *Times-Picayune*. Yet Törnqvist wasn't making a case for giving up. Projects like the diversions are urgently needed, he maintained, though such efforts might delay the retreat from the coast by only a generation or two.

Throughout the twentieth century, landscapes across the Mississippi River Basin were altered for the sake of benefits that, in some cases, never materialized or have disappeared. In hindsight, the intractable problems created by Gavins Point Dam on the Missouri and Fergie's fixes to the Mississippi probably weren't worth what the nation has paid. In each case, before such perilous alterations to the rivers were made, the long-term outcomes should have been more thoroughly considered, separate from the biases and incentives that too often motivate both proponents and opponents of a project. The post-1927 levees along the Mississippi might have been a better deal—protecting people against floods by depriving coastal Louisiana of sediment: a perpetual benefit for a perpetual cost. With man-made sediment diversions, governments are, perhaps for the first time, embracing sediment as part of a river management plan. Experts like Kolker and Wamsley can determine where sediment will accumulate and in what quantities, but they can't say whether their conclusions are good or bad. Rivers are live entities, ever dynamic—this is certain—but should their tendencies be discouraged or encouraged? Since stasis is not an option, which change is the better change?

"Who wants sediment is a values question, not necessarily a science question," Kolker said. Values are not scientific, but political.

Retreat and Fortify

Hurricane Katrina devastated Louisiana, ripping apart its surface and drowning its people, but deep down—ideologically—the storm didn't alter the way Americans live with water.

Though not a drop of the storm's rain crossed the Atlantic, the Netherlands was the place Katrina changed fundamentally. In August 2005, the Dutch stared at their televisions in horror as the streets of New Orleans flooded. Older citizens recalled the North Sea Flood of 1953, when eighteen hundred people died and parts of Rotterdam resembled storm-drenched New Orleans.

The Netherlands has a lot in common with south Louisiana. A small country laced with rivers and facing the sea, a third of its land sits below sea level and another third is vulnerable to flooding. Most of it occupies the delta of the Rhine and Meuse Rivers. For almost a thousand years, residents have depended on levees. To prevent catastrophe, the modern Netherlands (literally, *nether*, or low, lands) maintains more than ten thousand miles of man-made flood protection.

A comprehensive flood control plan had been proposed in the 1930s, but the Great Depression and World War II intervened. The flood of 1953 was the Dutch people's "never again" moment. (In Dutch, it is called *Watersnoodramp*, literally translated as "Waters-distress-disaster.") In response, their government created the first delta commission and began building the Delta Works, a massive network of walls, levees, dams, and moveable gates designed to hold back the sea. The American Society of Civil Engineers called the system, largely completed by the mid-1980s, one of the seven wonders of the modern world.

As Katrina spun away from New Orleans and the sun emerged, the Dutch should have felt reassured. They had the best flood defenses in the world, yet water experts there were worried. A disaster was still possible, they realized; absolute safety did not exist. The storm had caught America and its Corps of Engineers off guard. The Netherlands could not afford such an oversight.

"Katrina shook everyone awake," said a spokesman for the Dutch Traffic and Water Ministry three months after the storm. The ministry commissioned a new report, asking whether the Netherlands was prepared for such an event—could they withstand the rising seas and rushing rivers of the future? The Delta Works had been designed to hold back a storm surge of a certain height. The levees on the Rhine and Meuse Rivers were likewise designed to contain a certain volume of water flowing in from Germany and Belgium. Now the ministry feared that sea level rise and increasing precipitation had voided these assumptions.

The second delta commission presented its program in 2008. It revealed that if the Netherlands planned to survive for another century or two, it needed to do more. The Delta Works were not enough. In addition to strengthening floodwalls and levees, the Dutch people needed to proactively retreat from vulnerable areas. Only then would they be adequately protected. The Dutch had been devoted to levees since the fourteenth century, walling off delta land, pumping it dry, then living and farming on it. The commission was now asking them to reimagine, if not their national identity, then at least their flood control philosophy. It was asking them to give some of that hard-won land back.

This initiative, called Room for the River, had been in the works since the late 1990s; now it was incorporated into the Delta Program. In thirty-four places, people were relocated, levees were moved back, and the rivers were free to sweep in. One hundred and fifty houses and forty businesses were displaced and hundreds of acres of private land were taken. Unlike floodway and backwater areas on the Mississippi, almost all Dutch water storage areas and river bypasses operate passively. No one dynamites anything. As the rivers rise, water moves "naturally" into man-made auxiliary channels. In some places, the Dutch even degraded the level of farmland or carved new riverbeds, creating more *nether* land for the water to fill.

Before the government could implement Room for the River, it had to sell it to the Dutch people. "Planning for resilience with its inherent uncertainties is almost always a source of conflicts and politicization," wrote Delft University of Technology professor Hans de Bruijn in a 2015 paper. Since a river basin can encompass more than one town, county, or state, the beneficiaries of a flood protection project are often geographically distant from those who pay the costs. Think of the Birds Point–New Madrid Floodway, inundating Missouri to save Illinois. It's a win-lose game—some are sacrificed, while others stay dry.

To address this disparity, the Dutch minister broadened the discussion of local compensation to include parks, tourism, and even floating homes. Room for the River had two stated objectives: "Flood Protection" and "Spatial Quality." The minister created a website where citizens could propose and test various flood-mitigating scenarios, like lowering a levee to make a park, or putting houses on stilts. The people were listened to, and their requests were taken seriously. In exchange for giving up land, one town got a new park with a music venue, another got a deeper harbor. In this way, "a critical mass of 'winners' was created— parties who had emerged from the multi-game issue with an attractive package to show for it," wrote de Bruijn. Together, these "winners" formed a sense of the public good. Then the government could confront the holdouts; they could take the necessary land and forcibly move resisters. With the majority of citizens now behind them, politicians could weather the backlash.

Before Katrina, the Netherlands had budgeted €1 billion annually for water defense projects over twenty years. Another €500 million was spent annually on operations and maintenance. While Room for the River cost €2.3 billion and is almost complete, the entire suite of post-Katrina measures is projected to cost more than €1 billion extra per year through 2100. In return for their land and money, the Dutch people will receive flood protection that far surpasses anything in the United States, where most levees guard against only a 100-year flood (which has a 1-in-100 chance of occurring in any given year). Room for the River offered protection ranging from 1,250-year to 10,000-year. Most Americans would never consider such an unlikely scenario. The Dutch had no other choice. Their nation was at stake.

To explain why Room for the River had succeeded, de Bruijn's paper focused on two concepts: resilience and anticipation. "Resilience assumes that disturbances occur, but can be absorbed or 'controlled,'" he wrote. It also assumes that stabilization is possible after a catastrophe has receded. Anticipation, the antithesis of resilience, tries to stop specific threats by foreseeing them. Anticipation works well when threats can be accurately predicted, wrote de Bruijn, but "resilience is more effective as uncertainty, dynamics, and volatility increase."

Anticipation is the ideology of manifest destiny, of "fill the earth and subdue it." Climate change is uncertain, dynamic, and volatile, yet American planners are still determined to outsmart nature as if the planet were a chessboard. Engineers build something big and expensive, and tell the public they're protected. Then a Hurricane Katrina or a Missouri Valley Flood comes along and tears their expensive defenses apart. The public feels betrayed. The engineers excuse themselves, saying, "Sorry. We didn't anticipate that, so we didn't plan for it."

Several American cities—including Miami and New York—are almost as vulnerable to flooding as New Orleans. Yet no one has built much of anything there to guard against the water that most scientists say will come. The idea of Room for the River hinges on the fact that a wider floodplain equals lower flood crests. The U.S. Army Corps of Engineers knows this in theory. But, in most cases, politics has prevented the Corps from practicing it.

America's problem is more than just technical—it's cultural. Katrina inspired the Dutch to rethink their relationship with nature. It inspired the United States to build higher walls.

Katrina spared Lafourche Parish, a delta community perched perilously on sinking marshland, southwest of New Orleans. Windell

Curole manages the South Lafourche Levee District. Curole, a Cajun philosopher-poet, is quintessentially American—a migrant, a capitalist, and a romantic who doesn't want Big Government telling him what to do. His libertarian approach has proven effective in a deep red part of a red state. Without mentioning climate change, he simply says, "Without flood protection, nothing else matters."

His community has already answered several tough questions that the Dutch have faced but that the United States has not, such as: How much does comprehensive flood protection cost? Who will pay for it? Which areas can be saved, and which cannot?

Curole visited the Netherlands. They flooded for five hundred years, he said, and didn't take the problem seriously until 1953. However, from then until they finished building the Delta Works in 1998, the Dutch spent on flood protection the same percentage of their GDP that the United States spent on defense. "They pay 40 percent taxes and they don't bitch," Curole said. "The taxes are spent in a way that helps everybody." The Netherlands is a third the size of Louisiana, with almost four times the population and four times the GDP; its politics are extremely different, and the stakes there are as high as they can be, but for Curole, the underlying calculus is the same. "Every delta has great opportunity and great risk; you need to use that opportunity to balance the risk, or you flood."

The Dutch economy depends on trade. The sea that threatens the Dutch people also makes them rich. Like Curole, the Dutch realized that flood protection must bolster commerce, not restrict it. Rotterdam, in the heart of the Dutch delta, was the busiest port in the world for decades, and today it's the busiest outside Asia. The stretch of the Mississippi from Baton Rouge to New Orleans is the most vital commercial waterway in the United States; the Gulf Coast has also yielded tremendous wealth in oil and gas over the last century, but its flood defenses haven't benefited from the boom.

Most of Louisiana's oil industry is offshore now, in federal waters. Offshore tax revenues have traditionally bypassed Baton Rouge and gone straight to Washington. (Amos Cormier called this "the greatest wealth transfer in U.S. history.") In 2007, the state began receiving 37.5 percent of what the feds collect—up from zero—yet 90 percent of the offshore industry is based in, or supplied from, Louisiana. The state currently spends its share entirely on flood protection and coastal restoration, but it's not enough, said Curole. Every state except Alaska receives 50 percent of the revenue from onshore mining and drilling on federal land within its boundaries, and can spend those funds however it pleases. Similarly, Curole wants the feds to share a lot more of the offshore oil money with Gulf Coast states, where environments have suffered from nearly a century of extraction. He also fears that ecologically minded reformers will try to stifle the economy in the name of environmental restoration. That won't work, he is certain. Politicians should think, instead, about encouraging industry, while asking it to pay more to rebuild and maintain the landscapes it benefits from.

As their land subsided, Curole and the parish government realized—as the Dutch had—that in order to maintain their way of life in a changing world, they needed more money. The people of South Lafourche understood. They voted—73 percent in favor—to double their millage, or property tax rate. Ten years later, they approved a 1-cent sales tax increase by an 82 percent margin. These new revenue streams were dedicated solely to flood protection.

When the parish was planning the South Lafourche levee system in the late 1960s, its leaders had to decide: Who could they afford to protect? They chose to encircle four towns, lined up like beads on the thread of Bayou Lafourche: Larose, Cut Off, Galliano, and Golden Meadow. When a Tabby Cat Food plant proposed opening beyond Golden Meadow's corporate limit, the parish extended the levee to take in the plant and sixty additional houses. They left out the town of Leeville, where Curole's grandparents grew up, six miles closer to the Gulf.

"We said, 'Guys, this is where the line is,'" Curole recalled. "And the people on the other side hated us and we said, 'I'm sorry but this is it, or we do nothing.'"

Louisiana's governor appoints the board of the South Lafourche Levee District, and for thirty-eight years the board has employed Curole as managing director. "It's hard to do what we do and be elected. It takes a lot of guts to make the decisions we have. Most elected officials won't," he said. "We're not afraid."

In terms of flood protection, doing something is always better than doing nothing, Curole said, but too often American politicians choose to do nothing. They fear being disliked, delivering bad news, or taking someone's land, even if such measures will help protect an entire community. Politicians who promise that everyone will be able to rebuild after a flood tend to win elections. Those who exercise eminent domain or approve buyouts to move people out of high-risk areas tend to lose.

The South Lafourche levee system is simple: two gates that close off the upper and lower ends of Bayou Lafourche, connected by two arcs of levee. Inside, pumps suck out the ponding rain, typically 60 inches per year. With the gates closed, the area is protected from storm surges coming up the bayou or across the marsh. When the system went into operation, in 1986, Curole closed the gates maybe 20 days a year, for hurricane-induced high tides. In 2018, the gates were closed on more than 250 days. The land has sunk so much that now a slightly above average tide would put water in people's homes.

Most of South Lafourche makes its living by boat. If the gates are shut, only a few scheduled openings a day allow vessels in or out. If the tides rise too high, all openings cease, and the economy of the parish stands still, waiting for the water to fall. On a daily basis, Curole balances people's need to be protected from floods with their need to earn a living.

After Katrina, the Corps changed its definition of an "acceptable" levee, suddenly downgrading hundreds of systems that had met the old standards. The levees hadn't changed, but now many were considered "minimally acceptable" or "deficient." If a deficient levee failed, no one could blame the Corps. The cost to upgrade these levees to the newly stringent standards easily doubled the price of some projects. If the locals couldn't come up with their shares, the projects stalled. South Lafourche's Larose to Golden Meadow levee system was among those downgraded. Kicked out of the federal program, it was no longer eligible for government funds or engineering support. To re-qualify, the Corps told Curole, the district first had to pay for a five-year, $1 million study to find out where its levees were deficient. At the end of the study, there would be no guarantee of getting money to fix the deficiencies, especially if, given the higher price tag, the system now had a lower cost-benefit score.

Curole and parish leaders wanted to raise their levees. But, considering their budget and the new federal requirements, they decided to go it alone. They needed flood protection now. In lieu of a million-dollar study, Curole looked at historical Gulf Coast storm surges. The old levees stood thirteen feet high on the south end and eight feet high on the north. He decided the district could afford sixteen on the south and thirteen on the north, and handle all but the most extreme hurricanes.

Anyone who alters a federal project needs a permit, and the Corps refused to sanction Curole's levee upgrades. He went ahead and built them anyway. Every year, the Corps writes him up, calling the flood-gates and levee lifts "unauthorized construction." Curole ignores them. By setting these unreachable standards after Katrina, he believes, the Corps has actually made the Gulf Coast *less* safe than it would have been if Katrina had never happened. Curole will never forgive the Corps for the critical number of canceled or severely delayed flood protection projects.

In technical violation of federal law, South Lafourche began to raise its levees. First, the district needed more dirt and more rights-of-way. It sent letters to hundreds of landowners declaring its intention to take a slice of their property. Several families donated land, but a surge of lawsuits also swept into Curole's office. Fighting one class-action suit cost the district $3 million. Curole won, took the land, and raised the levee, but fifteen years later he was still incensed over the waste of time and money. In 1985, before the first iteration of the levee system was finished, the entire area flooded. "I didn't lose a minute of sleep, because I knew we had done everything we could," he said, but "I lose sleep over that lawsuit. It just eats me up, because we work so hard to spend every penny we got." Though the area hasn't flooded since '85, the new levees aren't finished, and as Curole sees it, his people have $3 million less protection because of those narrow-minded plaintiffs.

Eminent domain, taking private land in the public interest, has always stirred controversy. After Katrina, state senators in Louisiana worked to strengthen laws that classified hurricane and flood protection as public goods, worthy of eminent domain. Curole knew he stood on the right side of the post-Katrina laws, because he had helped draft them. Working closely with local legislators whose districts were even more vulnerable, he helped put the legal framework in place, without which, South Lafourche might be underwater today.

Yet even when Curole finally finishes his levee system, the result won't stand up to a storm like Katrina. The parish can't afford to protect against such long odds. "We can't guarantee you're not going to flood," said Curole. "I just want to be honest. I've got to work at telling people: Our flood protection, they think it's great, but they might think it's too great. A big enough storm gets over the top—we gone.

"If you recognize your risk, you reduce the chance of suffering that risk. If you ignore the risk, you increase the chance. And I'm telling you, our people understood the risk and feared it." People in South Lafourche have always worked in the marsh, beyond the levee. They watched the land sink. They saw how close they had come to annihilation. Most understood that they needed to pay more and get less in return.

Contrast this, said Curole, with the mindset of pre-Katrina New Orleans. The city's residents didn't spend time outside the levee, he said. They didn't know that things were changing out there; they assumed they were safe. "The Corps was not the reason New Orleans flooded. They were culpable, but they weren't the reason. The reason is people didn't fear flooding enough," Curole said. "They didn't want to face the risk. They didn't even talk about it."

Like the Dutch, the people of South Lafourche opted to retreat from indefensible areas and fortify those they deemed vital. They had to give up something or lose everything. Personally, Curole is comfortable with retreat. His ancestors have always moved as the land and water they live on has changed. First, they emigrated from France to present-day Canada. Then, in the mid-1700s, they were kicked out of Canada with the rest of the Acadians during *Le Grand Dérangement*. More recently, seven out of his eight great-grandparents fled the barrier island of Cheniere Caminada after the 1893 hurricane. They moved to Leeville, twelve miles inland. After the hurricane of 1915 destroyed much of Leeville, three out of his four grandparents moved again, farther up the bayou to higher ground. The 1915 storm surge reached just south of Golden Meadow, where Curole's mother was born. His father was born two towns up, in Cut Off, where the couple settled and where their son was born.

The locals know and trust Curole, a French-speaking Cajun like many older people in the parish. He wields power carefully. No bureaucrat or engineer sent down from Baton Rouge or Washington could have succeeded as he has.

Since the 2016 election, the word "infrastructure" has almost always been preceded by adjectives such as "aging," "decrepit," and "crumbling." These phrases have become bywords, repeated by politicians and the media so often that an average news consumer could be forgiven for thinking that every bridge they drive over is about to fall down, and that most municipal drinking water is laced with lead.

All infrastructure ages, and much of America's infrastructure has surpassed the length of time it was designed to last. Some is, indeed, crumbling, but most of the talk seeks merely to reset the clock, to rebuild everything just as it was, if not bigger and stronger. Moving dirt and pouring concrete benefits the politicians who want to bring pork home to their districts; it benefits civil engineers and construction companies. It does not necessarily benefit taxpayers or the planet.

Returning to the glory days of American civil works, the New Deal 1930s or the Interstate 1960s, would be a mistake, especially along the waterways, because we no longer live in that America. Back then, we did not understand the effects of man-made climate change; coal was our number one commodity; we had no Endangered Species Act, Clean Water Act, or Environmental Protection Agency; and we did little when poor, Black, and indigenous people were flooded, evicted, or left to live in high-risk areas. Before America rebuilds, America needs to rethink its infrastructure ideology. Without long-term planning—liberated from politics, special interests, and old assumptions—a New New Deal (or a Green New Deal) will be a colossal waste of resources. Americans are rarely willing to give up something in the near term for a benefit in the long term; our economic models aren't built this way, nor is our political process. The United States, as a country, is so young. Another century feels like forever.

The polder, a piece of reclaimed delta pumped out and surrounded by levees, is the most basic unit of Dutch government. Long before the modern Dutch state, in the Middle Ages, disparate users of these polders began working together to keep their land dry. The polder model is characterized by consultation, consensus, and compromise, and it established the social foundation that made Room for the River possible. A polder is defined by topography and the way water drains, not arbitrary borders. It is the logical administrative unit for planning flood defenses. Basins, also defined by water, not politics, are the logical units for river management because interests within a basin coincide; water coming over a dam in St. Paul, Minnesota, will soon arrive in Memphis, Tennessee.

The only basin-based governance models in the United States are the Tennessee Valley Authority and the Mississippi River Commission, both of which have proven very effective at large-region water management. The diverse groups that the Mississippi River Commission oversees—at least on the lower river—all feel heard. They have a direct line of communication with the generals and engineers who govern them. When the commission must sacrifice something for the greater good, it can garner enough support within the basin to do so. The commission also occupies a crucial middle ground between regional interests and the distant bureaucracy of Washington. With three civilian members and one from NOAA, the seven-member body can be free (and should be freer) to disagree with special interests, locals, and D.C. What if there were more commissions—for the Colorado and Columbia Rivers, the Chesapeake Bay, the Sacramento–San Joaquin Delta, and dozens of other watersheds? These bodies could be funded and empowered to act before climate change or the pace of infrastructural decay forces desperate ad hoc measures. Passing out multibillion-dollar aid packages after each disaster is wasteful and backward. The purpose of planning is to solve problems in advance, not make decisions in the teeth of a crisis.

Perhaps more important than what the Dutch have built is the organizational, financial, and legal framework that makes such projects possible. In 1950 there were 2,650 entities owning and maintaining flood defenses in the Netherlands; now there are twenty-one water boards overseeing every aspect of ninety-five different ring levees. Those inside the rings pay taxes and vote on policy, and the water boards report to the twelve provinces, which report to the central government. For centuries, the Dutch approach to water has been pragmatic and holistic. Dutch water boards are responsible not just for preventing floods, but for water quality and wastewater treatment. Article 21 of the Dutch Constitution states: "Government care is aimed at the habitability of the country and the protection and improvement of the environment." In the Netherlands, habitability has always been synonymous with flood control. The Dutch also regularly reassess their infrastructure and adjust their laws to adapt to new science and a changing world, and they spend a lot of time and money planning for a crisis. *What if* the 10,000-year levees overtop?

A basin-based approach supported with an adequate legal framework may be the only way to garner enough public support to reimagine America's rivers. Like Windell Curole, the Mississippi River Commission has the authority and trust to make sacrifices palatable. If the commission president told constituents, "Look, we need to give this land up or we're all going to flood," they would listen, and most would believe him or her. Then the commission could start negotiating, and giving and taking, until it had satisfied a critical mass of winners. It could then confront the holdouts and make the necessary changes—constructing here, deconstructing there—to protect people for another century.

Americans love stories about "saving" places like Leeville, where Curole's parents grew up. We don't have a similar narrative for leaving places. Maybe, if humans make a monumental commitment to changing their behavior and their economies, we can slow down climate change, and the sea won't rise so high, and the rain won't fall so hard. It is possible to reverse global warming. But even if the whole planet stopped emitting heat-trapping gasses today, the earth would continue to warm for a few decades before its temperature stabilized and eventually began to decline. In actuality, experts predict that by the end of this century, global sea level will rise twelve inches and much of the United States will see a 20 to 40 percent increase in extreme precipitation. And this is a best-case scenario. The sea will rise. It will rain more. Unless we take decisive and preemptive action to protect ourselves, floods like 2011 and storms like Katrina—or worse—will overcome our defenses. America can buy a lot more risk reduction, but it cannot buy down every risk. We need to imagine a prosperous country that has both retreated and fortified. Daunting as the task may seem, we don't have a choice.

Acknowledgments

A friend and I were put-putting down the Ohio River in a sixteen-foot aluminum boat, camping on the bank at night. On the afternoon of what we thought was our last day, the outboard motor broke down just past Lock and Dam No. 53. A worker there didn't like the look of us rowing clumsily toward the Olmsted construction site, so he hopped in a skiff and towed us back to the lock. I ended up cooking dinner in the pump house with Mike Burton, who was on the night shift. He shared his bread and stories with me, and I watched him fire up the maneuver boat's massive boiler. I slept on the bench seat of the lock's little towboat.

The next morning, I met Randy Robertson, No. 53's lockmaster. He asked two of his mechanics, Kenny Robertson and Bryan Parrett, to look at our motor. After scrubbing the sparkplugs and adding something to the fuel to counteract the watery Kentucky gas, we were on our way. We passed the awesome spectacle of Olmsted, then about two-thirds complete, and proceeded to Cairo, the last stop on our journey.

I had glimpsed a world, and I couldn't stop thinking about it.

Now, here I am, six years later, at the end of a book about America's rivers and the people who tame them or are tamed by them. First, I would like to thank everyone who appears in this book and who trusted me with their stories, especially Randy Robertson, Luther Helland,

Lester Goodin, and Twan Robinson. I also want to thank a select group who, in addition to being experts in their fields, know a lot of other knowledgeable people: Charles Camillo, Deb Calhoun, and Sandy Stockholm each set me up with a web of excellent sources and resources. I was very fortunate to encounter the Mississippi River Commission under the leadership of Major General Richard Kaiser, who included me in candid conversations and meetings, and introduced me to several people who became pivotal to this book, including Ty Wamsley. There are also a few people who explained things to me in the very beginning, when I was just starting to grasp how big these issues really are: T. Stephen Gambrell, W. Dustin Boatwright, and Kenneth R. Olson.

I want to credit a group of individuals whose names do not all appear in the text of the book, but who helped with the tremendous job of fact-checking it. Thank you: Brian Rentfro, John O. Anfinson, Todd Shallat, Rod Lincoln, Robert Kelley Schneiders, James F. Barnett Jr., Roy Van Arsdale, Patrick Mulvany, Richard Campanella, Kent Parrish, Robert Slomp, Ken Eriksen, David Blalock, and Torbjörn E. Törnqvist.

On the writing side are my two mentors (who tend to give the opposite advice): Barry Newman and Susan Shapiro. I can't begin to list all the things I wouldn't know if it weren't for these two. I also want to thank a few dedicated readers and supporters of this project: Jim Jennewein, Frank Flaherty, all the members of Sue's Sunday Writing Group, and my father, Jeff Kelley.

Then there is my agent, Renée Zuckerbrot, a good listener and shrewd strategist who always calls me back; my superb editor at Avid Reader Press, Jofie Ferrari-Adler, who helped define and shape this project and was very patient with me; and Jofie's lieutenant, the ever-kind and indispensable Carolyn Kelly. Also Jesse Pesta at the *New York Times*, who took a chance on an article about an old dam, and later helped put another dam story on the front page. And I may well owe my career to Jill Kirschenbaum, formerly of the *Wall Street Journal*, who told me the secret to pitching an A-Hed.

Finally, thank you to Araby Kelley, my wise and untiring wife, and to my overwhelmingly cool son, Anders, who has lived his entire life in the shadow of this project.

Notes

xi **Prelude: Perfecting Nature:** This chapter title is borrowed from Todd Shallat, *Hope for the Dammed: The U.S. Army Corps of Engineers and the Greening of the Mississippi* (Boise State University: Faculty Authored Books, Book 391, 2014), http://scholarworks.boisestate.edu/fac_books/391.

xi **An unprecedented amount of snow and rain:** Sixty-one million acre-feet of water would flow into the Missouri Basin that year, enough to cover New York, New Jersey, Connecticut, Vermont, and New Hampshire with more than a foot of water.

xiv **The river was doing what:** Robert Kelley Schneiders, *Unruly River: Two Centuries of Change Along the Missouri* (Lawrence: University Press of Kansas, 1999). Schneiders told me that the Missouri excavated its present course during the Wisconsin Glaciation Episode, which lasted from 75,000 BP to about 11,000 BP.

xiv **When Lewis and Clark traveled:** I used the Journals of the Lewis and Clark Expedition Online, the text of the Nebraska edition of the Lewis and Clark journals, edited by Gary E. Moulton, and hosted by the University of Nebraska, Lincoln (https://lewisandclarkjournals.unl.edu). I used the original spelling, grammar, and punctuation of the journals wherever possible.

xiv **The men shot two deer:** John Ordway, a member of the Corps of Discovery, wrote in his journal of July 19, 1804: "We gethered a quantity of cherries at noon time & put in to the Whisky barrel."

xiv **Clark concluded that evening's:** A portion of William Clark's entry from the same day: "As we approach this Great River Platt the Sand bars are much more noumerous than they were, and the quick & roleing Sands much more danjerous, where the Praries aproach the river it is verry wide, the banks of those Plains being much easier to undermine and fall than the wood land passed a willow Island Situated near the middle of the river, a Sand bar on the S. S. and a Deep bend to the L S. camped on the right Side of the Willow Island—W. Bratten hunting on the L. S Swam to the Island. Hunters Drewyer killed 2 Deer, Saw great numbers of young gees. The river Still falling a little Sand bars thick always in view."

xiv **In places, the river:** Joseph Whitehouse, a member of the Corps of Discovery, wrote in his journal of June 2, 1804: "The Mesouri River at this place is 875 Yards wide."

xv **The purposes of Lewis and Clark's expedition:** See Thomas Jefferson's instructions to Meriwether Lewis, June 20, 1803 ("The object of your mission is to explore the Missouri river, & such principal stream of it, as, by it's course & communication with the waters of the Pacific ocean may offer the most direct & practicable water communication across this continent, for the purposes of commerce.") and Jefferson's January 18, 1803, letter to Congress asking for authorization and funding for the expedition ("In order peaceably to counteract this policy of theirs, and to provide an extension of territory which the rapid increase of our numbers will call for, two measures are deemed expedient. First: to encourage them to abandon hunting, to apply to the raising stock, to agriculture and domestic manufacture, and thereby prove to themselves that less land and labor will maintain them in this, better than in their former mode of living. The extensive forests necessary in the hunting life, will then become useless, and they will see advantage in exchanging them for the means of improving their farms, and of increasing their domestic comforts. Secondly: to multiply trading houses among them . . .") Both available at Monticello.org.

xvi **While the roads and rails may carry more goods by volume:** "Importance of Inland Waterways to U.S. Agriculture," prepared for the USDA Agricultural Marketing Service by Informa Agribusiness Consulting, August 2019.

xvi **To protect people and property:** See the National Levee Database: https://levees.sec.usace.army.mil/#/. China's Great Wall is, depending on which walls are included in the estimate, about thirteen thousand miles long.

xvi **Since 1928, taxpayers have spent:** Personal communication with Brian Rentfro, the U.S. Army Corps of Engineers' Mississippi Valley Division historian. Rentfro fact-checked a good portion of my book, especially the sections dealing with the Mississippi River Commission, the MR&T Project, and the Great Flood of 1927.

xvii **Donald Trump said Clinton's pledge was:** Melanie Zanona, "Trump Says He'd Double Clinton's $275 Billion Infrastructure Plan," *The Hill*, August 2, 2016, https://thehill.com/policy/transportation/290121-trump -on-paying-for-infrastructure-projects-well-get-a-fund.

xvii **In his Inaugural Address, he said the country:** *Time* staff, "Read Donald Trump's Full Inauguration Speech," *Time*, January 20, 2017, https://time .com/4640707/donald-trump-inauguration-speech-transcript/.

xvii **But, in May, Trump walked out:** Tim Hains, "Pelosi After Trump Walked Out of Infrastructure Meeting: 'I Pray for the President' and the Country," *RealClear Politics*, May 22, 2019; Phillip Ewing, "Trump Scorches Democrats as Pelosi Broaches Prospect of 'Impeachable Offense,'" NPR, May 22, 2019.

xviii **The quality of American infrastructure was ranked tenth:** James McBride, "The State of U.S. Infrastructure," Council on Foreign Relations, January 12, 2018, https://www.cfr.org/backgrounder/state-us-infrastructure.

xviii **And U.S. infrastructure dollars were increasingly going toward:** Joseph W. Kane and Adie Tomer, "Shifting into an Era of Repair: US Infrastructure Spending Trends," Brookings.edu, May 10, 2019, https://www.brookings.edu /research/shifting-into-an-era-of-repair-us-infrastructure-spending-trends/.

xviii **In 2017, the American Society of Civil Engineers:** "America's Infrastructure Report Card 2017," published by the American Society of Civil Engineers (ASCE). See separate sections for "Inland Waterways," "Dams," and "Levees," https://www.infrastructurereportcard.org/.

xviii **When Trump spoke about "inland waterways":** According to Waterways Council, Inc., Obama mentioned inland waterways as a candidate, but never as president.

xviii **More than a third of levees:** According to the National Levee Database (https://levees.sec.usace.army.mil/), accessed on July 26, 2020: Out of 28,000 total levee miles in the U.S., 15,400 miles are rated "Unacceptable" or "Minimally Acceptable," and 9,680 miles are rated "Unacceptable," 1,536 miles of which were constructed by and are operated by the Corps.

xix **The original plan for damming and diking the Missouri:** Roger S. Otstot, "An Overview of the Pick-Sloan Missouri Basin Program," Bureau of Reclamation, Great Plains Region; also personal communications with Robert Kelley Schneiders, author of *Unruly River*.

xix **Though this hodge-podge system:** Jeff Haldiman, "Progress Being Made on Missouri River Levee Repairs," *News Tribune* (Jefferson City, MO), June 25, 2020.

xx **After the flood of 2011 receded, engineers with the Corps' Omaha District:** "Assessment of Conceptual Nonstructural Alternative Levee Setbacks along the Missouri River (Lower L-575 / Upper L-550 and Lower L-550)," Final Report, April 2012.

xx **The local levee board would buy the strips of land:** Ibid.

xx **The Corps could have acquired the land:** "We have the authority to condemn the land—eminent domain," said John Remus, chief of Missouri River Basin Water Management for the Corps. "We're not going to go there. It's political suicide. We haven't even gone there and it's still kind of political suicide." Remus, whom I interviewed, was part of the delegation that proposed the plan to the Iowa farmers in 2011. One farmer still keeps a diagram that Remus left behind. To him, it demonstrates the Corps' intention to take his land.

xx **At the time, Army Corps of Engineers Colonel:** Ron Toland, "Army, USACE open doors to new Ansbach lodge," Army press release, December 19, 2011, https://www.army.mil/article/71027/army_usace_open _doors_to_new_ansbach_lodge.

xxi **The loveless shotgun wedding:** James Patton, president of the National Farmers Union (it supported a Missouri Valley Authority), called the Corps-Reclamation compromise "a shameful, loveless shotgun wedding," https://www.nps.gov/mnrr/learn/historyculture/pick-sloan-plan-part-two -debate-and-compromise.htm. A Missouri River Commission did exist from 1884 to 1902, https://www.cerc.usgs.gov/data/1894maps/.

xxii **Twain's book begins:** According to the Online Etymology Dictionary: "infrastructure (n.) 1887, from French *infrastructure* (1875); see infra- + structure (n.). The installations that form the basis for any operation or system. Originally in a military sense."

xxii **Many of Lewis and Clark's riverbank campsites:** Robert Criss, "Mid-Continental Magnetic Declination: A 200-Year Record Starting with Lewis and Clark," *GSA Today*, October 2003.

xxii **Steamboats have been unearthed:** See the Arabia Steamboat Museum in Kansas City. The *Arabia* sank in 1856 and was discovered in 1988 "45 feet underground and a half-mile away from the present channel of the Missouri River," https://www.1856.com/arabia-story.

xxiii **depicted the river's meanderings over six thousand years:** H. N. Fisk, "Geological Investigation of the Alluvial Valley of the Lower Mississippi River," U.S. Army Corps of Engineers, 1944, 37.

xxiv **Kaiser was referring to a report:** Bill Frederick, "Precipitation Trends in the Mississippi River Watershed," U.S. Army Corps of Engineers, Mississippi Valley Division / National Weather Service, February 2019.

xxiv **Bonnet Carré had been opened in three of the previous four years:**
Bonnet Carré was also opened in April 2020, for a new record of four out
of the past five years.

PART I—The Lock

9 **The captain of the *Exxon Valdez*:** Exxon Valdez Oil Spill Trustee Council, Questions and Answers about the Spill, https://evostc.state.ak.us
/oil-spill-facts/q-and-a/.

10 **The owners may want them to push on:** Of a steamboat pilot—the
equivalent of a modern captain—Twain wrote in *Life on the Mississippi* that
his "pride in his occupation surpasses the pride of kings."

12 **Compared to roads, where the average American:** W. Kim, V. Añorve,
and B. C. Tefft, "American Driving Survey, 2014–2017," AAA Foundation for Traffic Safety, 2019, https://aaafoundation.org/american-driving
-survey-2014-2017/.

13 **Supposed I had gotten over Logtown riffle:** Lewis's journal of September 2, 1803, original spelling and grammar, https://lewisandclarkjournals.unl
.edu/journals. Length of the boat, duration of journey to the Mississippi,
etc., all from the journals and the UNL site.

14 **Timothy Flint, a Presbyterian minister from Massachusetts:** John Ervin
Kirkpatrick, *Timothy Flint, Pioneer, Missionary, Author, Editor, 1780–1840:
The Story of His Life Among the Pioneers and Frontiersmen in the Ohio and Mis-
sissippi Valley and in New England and the South* (Cleveland: The Arthur H.
Clark Company, 1911), 18, 21, 31.

14 **There he encountered teamsters:** Timothy Flint, *Recollections of the Last
Ten Years, Passed in Occasional Residences and Journeyings in the Valley of the
Mississippi, from Pittsburg and the Missouri to the Gulf of Mexico, and from
Florida to the Spanish Frontier; in a Series of Letters to the Rev. James Flint, of
Salem, Massachusetts* (Boston: Cummings, Hilliard, and Company, 1826), 8.

14 **Ohio was considered "back woods":** Ibid., 11, 28.

14 **in Pittsburgh, Flint noted the "funereal" atmosphere:** Ibid., 18.

14 **Eventually, Flint booked passage:** Ibid., 18–20.

15 **A farmer, Flint imagined:** Ibid., 29.

15 **He passed towns of thousands:** Ibid., 22.

15 **while glorying in:** Ibid., 27.

15 **Four years later, the United States government:** Leland R. Johnson,
*Men, Mountains and Rivers: An Illustrated History of the Huntington District; US
Army Corps of Engineers, 1754–1974* (Washington, DC: U.S. Government
Printing Office, 1977), 17. I am indebted to the works of the late Leland

Johnson, who produced dozens of fact-filled and engaging histories for the Corps, often with the help of Charles E. Parrish.

15 **The *Western Engineer* drew only nineteen inches of water:** Ibid., 18.

15 **Edwin James, botanist and geologist:** Edwin James, *Account of an Expedition from Pittsburgh to the Rocky Mountains, Performed in the Years 1819 and '20, By Order of Hon. J. C. Calhoun, Sec'y of War: Under the Command of Major Stephen H. Long, from the Notes of Major Long, Mr. T. Say, and Other Gentlemen of the Exploring Party* (Philadelphia: H. C. Carey and I. Lea, 1823), 19.

16 **At Wheeling, Virginia (now West Virginia):** Ibid., 17; Dolores N. Savage, "The Relation of the Federal and State Governments to the Problem of Internal Improvements, 1810 to 1840" (Loyola University Chicago, Master's Theses 357, 1935). The national road eventually crossed the Ohio and continued to Vandalia, Illinois, sixty-five miles shy of its goal: St. Louis and the Mississippi. The total cost was $7 million.

16 **The defining impediment to navigation on the Ohio:** James, *Account of an Expedition*, 26. See also: Leland R. Johnson and Charles E. Parrish, *Triumph at the Falls: The Louisville and Portland Canal* (Louisville: Louisville District, U.S. Army Corps of Engineers, 2007), 74. The limestone ledges were eventually blasted away to aid navigation and quarried for cement.

16 **Influenced by the British tradition of dead-water canals:** Todd Shallat, *Structures in the Stream: Water, Science, and the Rise of the U.S. Army Corps of Engineers* (Austin: University of Texas Press, 1994), 39.

17 **George Washington envisioned a canal crossing:** Harry Sinclair Drago, *Canal Days in America: The History and Romance of Old Towpaths and Waterways* (New York: Clarkson N. Potter, Inc., 1972), 47–71.

17 **The closest the canal companies came:** Ibid., 143-148.

17 **The first commercially viable steamboat:** Johnson, *Men, Mountains*, 19.

18 **Fulton wanted to make himself great, too:** Ibid., 19–20.

18 **a Supreme Court ruling broke the Fulton monopoly in 1824:** The case was *Gibbons v. Ogden*. See: http://www.phschool.com/curriculum_support /interactive_constitution/scc/scc12.htm.

18 **A private corporation chartered by the Kentucky legislature:** Drago, *Canal Days in America,* 251–254. See also: Rick Lee Woten, "Navigating internal improvement: rivers, canals, and state formation in the nineteenth-century midwest" (Iowa State University Graduate Theses and Dissertations 10755, 2009).

18 **A week after taking command of the Continental Army:** Shallat, *Structures in the Stream*, 30, and personal communications.

19 **sent a bill to President James Monroe expanding the U.S. Army Corps of Engineers' authority:** Johnson, *Men, Mountains*, 25.

19 The army engineers were among the young nation's few experts: Shallat, *Structures in the Stream*, and personal communications.

19 Still within the army, but with a mostly civilian workforce: Online biography of Lieutenant General Todd T. Semonite: fifty-fourth chief of engineers and commanding general of the U.S. Army Corps of Engineers, published by usace.army.mil on June 5, 2018.

19 Before it took over the canal around the Great Falls: Shallat, *Structures in the Stream*, 101, and personal communications.

19 The dike diverted the river's flow: Johnson, *Men, Mountains*, 27.

19 also began a campaign to remove snags: Ibid., 27–38.

19 "as neatly as if it were a sliver in your hand": Mark Twain, *Life on the Mississippi* (Boston: James R. Osgood and Company, 1883). I used the Project Gutenberg EBook: EBook # 245, August 20, 2006, http://www .gutenberg.org/files/245/245-h/245-h.htm.

20 Snags were so common and so reviled: Johnson, *Men, Mountains*, 25.

20 "Service on Uncle Sam's toothpullers": Ibid., 41.

20 Within a few decades of Fulton's: Ibid., 33.

20 "I was compelled to know, to my cost": Flint, *Recollections*, 25.

20 By flatboat, the trip from New Orleans: "A History of Steamboats," USACE Mobile District, https://www.sam.usace.army.mil/Portals/46/docs /recreation/OP-CO/montgomery/pdfs/10thand11th/ahistoryofsteamboats .pdf. In 1870 the steamboat *Natchez* raced the *Robert E. Lee* from New Orleans to St. Louis. The *Robert E. Lee* won the race, arriving in St. Louis after three days. The *Natchez* arrived six hours later.

20 Floating down the Ohio on a beautiful spring morning: Flint, *Recollections*, 15.

21 Samuel Clemens was such a: The Mark Twain House & Museum, online biography page, https://marktwainhouse.org/about/mark-twain/biography/.

21 In the 1830s, there were twenty-three miles of track: "A History of Steamboats."

22 The last straw for Twain's gingerbread steamers: Twain, *Life on the Mississippi*.

23 Back in the 1840s, a series of primitive: Leland R. Johnson, *The Davis Island Lock and Dam, 1870–1922* (U.S. Army Engineer District, Pittsburgh, Pennsylvania, 1985), 9.

23 "During several months in most years, Pittsburgh's steamboats": Ibid., 2.

23 Army engineers had traveled: Ibid., 37.

24 The Davis Island Lock and Dam, completed in 1885: Shallat, *Structures in the Stream*, 191.

24 **Trucks moved six and a half times as much:** "Weight and Value of Freight Shipments by Domestic Mode: 2017," U.S. Department of Transportation, Bureau of Transportation Statistics and Federal Highway Administration, Freight Analysis Framework, version 4.5, 2019.

25 **the wicket design was scrapped:** Leland R. Johnson, *The Falls City Engineers: A History of the Louisville District, Corps of Engineers, United States Army* (U.S. Army Corps of Engineers, 1974), 242–243.

29 **The crane—called a "lever rack"—did not meet modern safety standards:** See federal labor law, 29 CFR § 1926.1426, a free fall boom is prohibited if "an employee is in the fall zone of the boom or load," which was the case every time a wicket was lifted, https://www.osha.gov/laws-regs /regulations/standardnumber/1926/1926.1426.

43 **The most definitive account of what happened next:** United States Government Accountability Office, *Factors Contributing to Cost Increases and Schedule Delays in the Olmsted Locks and Dam Project*, Report to Congressional Committees, February 2017, GAO-17-147.

44 **These studies, commissioned by the Corps:** Ibid. Various studies, commissioned or conducted by the Corps, found that in-the-wet would cost between $40 million and $60 million less, and save between one and a half and five and a half years.

45 **He waited until he had maximum leverage to push it through:** Washington Bureau, "Bill That Reopened Government Full of Pet Projects," *Atlanta Journal-Constitution*, October 17, 2013, https://www.ajc.com/news /national-govt—politics/bill-that-reopened-government-full-pet-projects /UGkKcC98ntkiXeIHoIqJrM/. See also: "Transcript of Pelosi Press Conference Today," October 17, 2013, press release, https://pelosi.house.gov /news/press-releases/transcript-of-pelosi-press-conference-today-52.

49 **To the subsidy averse, this still amounted to a huge giveaway:** Testimony of Steve Ellis, vice president, Taxpayers for Common Sense, Subcommittee on Water Resources and Environment, Committee on Transportation and Infrastructure hearing on "The Economic Importance and Financial Challenges of Recapitalizing the Nation's Inland Waterways," September 21, 2011, https://www.taxpayer.net/infrastructure/tcs -testifies-on-the-inland-waterways-trust-fund/.

49 **Steve Ellis, vice president of Taxpayers for Common Sense:** Ibid.

49 **"The atypical cost-sharing structure":** Ibid.

49 **Flood control, the Corps' primary mission along with navigation:** Personal communication with Kent Parrish, Mississippi River Levees project manager for the Corps' Vicksburg District.

51 **An outstanding example was the lengthy effort to:** "Congress Fattens

Up the Pork Barrel: Billions Thrown Away—and It's Your Money," *Life*, Vol. 55, No. 7, August 16, 1963.

52 **Then there's the J. Bennett Johnston Waterway:** Michael Grunwald, "A River in the Red: Channel Was Tamed for Barges That Never Came," *Washington Post*, January 9, 2000, www.washingtonpost.com/wp-srv /WPcap/2000-01/09/046r-010900-idx.html.

52 **Earmark spending reached:** CAGW Staff, "Time to End Earmarks Once and for All," The WasteWatcher, Citizens Against Government Waste, January 30, 2012.

53 **Since 1936, the Corps had weighed costs against benefits:** National Research Council, *Analytical Methods and Approaches for Water Resources Project Planning* (Washington, DC: The National Academies Press, 2004), 38, https://doi.org/10.17226/10973. "The Flood Control Act of 1936 mandated formal benefit-cost analysis (BCA) within Corps planning studies. One observer has referred to the act as '. . . one of the heroic efforts of the United States Congress to control its own bad habits' (Porter, 1996)."

53 **In 1998, a group of Corps economists completed an analysis:** Personal communication with Donald Sweeney II.

53 **The original study's managing economist:** Office of Special Counsel, Affidavit of Donald C. Sweeney, February 1, 2000. See also: OSC File No. DI-00-0792, letter dated December 6, 2000, for the Office's findings in Sweeney's favor.

53 **"If the demand curves, traffic growth projects and associated variables":** Michael Grunwald, "How Corps Turned Doubt into a Lock: In Agency Where the Answer Is 'Grow,' a Questionable Project Finds Support," *Washington Post*, February 13, 2000, A01.

53 **The inspector general's report vindicated Sweeney:** U.S. Army Inspector General, Report of Investigation, Case 00-019, 2000.

53 **The 2001 NRC report determined that:** National Research Council, *Inland Navigation System Planning: The Upper Mississippi River–Illinois Waterway* (Washington, DC: The National Academies Press, 2001), https://doi .org/10.17226/10072.

54 **According to a 2019 report commissioned by the U.S. Department of Agriculture:** "Importance of Inland Waterways to U.S. Agriculture," prepared for the USDA Agricultural Marketing Service by Informa Agribusiness Consulting, August 2019.

54 **the Corps told Congress that 113 million tons of cargo would transit:** GAO, *Factors Contributing to Cost Increases and Schedule Delays*.

55 **In 1975, the Corps had predicted:** National Research Council, *Inland Navigation System Planning*, 46.

55 **Chickamauga Lock and Dam, on the Tennessee River:** John Frittelli, "Prioritizing Waterway Lock Projects: Barge Traffic Changes," Congressional Research Service, updated June 1, 2018, R45211, and personal communication with Deb Calhoun.

56 **Toy explained that the Corps had come up with a process:** The idea is that if the benefits remain the same, but you've already spent half the money, the ratio will be higher if you apply the remaining cost to the original benefit.

56 **Now that Olmsted was finished, the towing industry:** Frittelli, "Prioritizing Waterway Lock Projects," and personal communication with Deb Calhoun.

57 **Each of Olmsted's 140 wickets:** As of fall 2020, the Olmsted crew could raise a wicket in five minutes.

PART II—Alluvial Empire

68 **A levee is usually built with a three-to-one slope:** Personal communication with Kent Parrish. The mainline Mississippi River levees are built even wider, with a five-to-one slope on the land side, and four-to-one on the river side.

70 **In 1927, when a levee burst at nearby Cabin Teele:** John M. Barry, *Rising Tide: The Great Mississippi Flood of 1927 and How It Changed America* (New York: Touchstone/ Simon & Schuster, 1997), 281.

70 **The Corps once commissioned a map of the river's previous channels:** H. N. Fisk, "Geological Investigation of the Alluvial Valley of the Lower Mississippi River," U.S. Army Corps of Engineers, 1944, 37.

73 **In Arkansas, a man was murdered in a dicamba dispute:** Marianne McCune, "A Pesticide, a Pigweed and a Farmer's Murder," *Morning Edition*, National Public Radio, June 14, 2017, https://www.npr.org/2017/06/14/5000879755/a-pesticide-a-pigweed-and-a-farmers-murder.

73 **Bayer, a German pharmaceutical company, purchased Monsanto:** Tim Loh, "Bayer CEO Opens Door to Roundup Settlement as Lawsuits Swell," *Bloomberg*, July 30, 2019, https://www.bloomberg.com/news/articles/2019-07-30/bayer-cautions-on-profit-as-roundup-plaintiffs-rise-to-18-400; Caroline Winter and Tim Loh, "With Each Roundup Verdict, Bayer's Monsanto Purchase Looks Worse," *Bloomberg*, September 19, 2019, https://www.bloomberg.com/news/features/2019-09-19/bayer-s-monsanto-purchase-looks-worse-with-each-roundup-verdict.

73 **In June 2020, Bayer agreed to settle about ninety-five thousand:** Eli

Chen, "Bayer To Pay Nearly $11 Billion To Settle Current And Future Roundup Cancer Suits," St. Louis Public Radio, June 24, 2020.

75 **Fifty percent of Americans worked in agriculture:** Ezra Klein and Susannah Locke, "40 maps that explain food in America," Vox.com, June 9, 2014, https://www.vox.com/a/explain-food-america (map source: USDA).

76 **"Like Faulkner says, You don't own the land":** The Faulkner quote, from *The Unvanquished*, is actually "[Uncle Buck and Uncle Buddy] believed that land did not belong to people but that people belonged to land and that the earth would permit them to live on and out of it and use it only so long as they behaved and that if they did not behave right, it would shake them off just like a dog getting rid of fleas." Located using Southeast Missouri State University's The Quotable Faulkner site, https://semo.edu/cfs/quotes.html.

76 **Though he was from the hills, Faulkner spent a good deal of time in the flatlands:** Phillip Gordon, "The Delta and Yoknapatawpha: The Layering of Geography and Myth in the Works of William Faulkner," *Study the South*, Center for the Study of Southern Culture at the University of Mississippi, November 28, 2016, https://southernstudies.olemiss.edu/study-the-south/delta-yoknapatawpha/.

76 **In his novel *The Wild Palms*:** William Faulkner, *The Wild Palms* (Middlesex, England: Penguin Books Ltd, 1939).

77 **To sense that drama, Goodin recommends *Rising Tide*:** Barry, *Rising Tide*, 14.

77 **Beginning in December of the previous year:** A. J. Henry, "Frankenfield on the 1927 Floods in the Mississippi Valley," *Monthly Weather Review*, December 3, 1927; Barry, *Rising Tide*, 194.

77 **New Orleans endured 14.96 inches in eighteen hours:** Barry, *Rising Tide*, 15.

77 **The first government levee to breach was at Dorena:** Henry, "Frankenfield on the 1927 Floods."

78 **But without the outlet created by that breach:** Ibid.

79 **Everyone in Greenville knew the river was rising:** Barry, *Rising Tide*, 198–201.

79 **A year after the flood, Lonnie Johnson released:** Lonnie Johnson, "Stay Out of Walnut Street Alley / Broken Levee Blues," Okeh Phonograph Corporation, Shellac, 10", 78 RPM, US, 1928.

79 **Five days after the Dorena breach:** Barry, *Rising Tide*, 200.

79 **"The roar of the crevasse drowned all sound":** Ibid., 202.

79 **The Mounds Landing crevasse inundated an area:** Ibid., 206.

80 **"There lay a flat still sheet of brown water":** Faulkner, *The Wild Palms,* 46.

80 **In an article published by the University of Mississippi:** Gordon, "The Delta and Yoknapatawpha."

80 **recovering more than 16.8 million acres:** Brian Rentfro said that this number was derived in 2011, after engineers completed a post-flood report comparing 2011 to 1927 using modern computer models and historical inundation maps.

81 **Garcilaso de la Vega, also known as El Inca:** Gloria A. Young and Michael P. Hoffman (eds.), *Expedition of Hernando de Soto West of the Mississippi, 1541–1543: Proceedings of the de Soto Symposia, 1988 and 1990* (University of Arkansas Press, 1999), 173.

81 **"Which in the beginning overflowed the wide level ground":** The English translation of this quote appears as the epigraph to the definitive report on the 1993 flood on the Upper Mississippi River. *Sharing the Challenge: Floodplain Management into the 21st Century*, report of the Interagency Floodplain Management Review Committee to the Administration Floodplain Management Task Force (Washington, DC, 1994).

82 **This process—common in the past:** Some river experts believe that cutoffs occurred with unnatural frequency in the 1800s. Riverbanks had been deforested to fuel steamboats and, with no trees to hold the banks in place, erosion accelerated, rivers moved around more, and their waters became muddier than at any other time in history. See: Brien R. Winkley, "Man-Made Cutoffs on the Lower Mississippi River, Conception, Construction, and River Response," U.S. Army Engineer District, Vicksburg, March 1977. Andy Schimpf also talked about this.

82 **During the Civil War, Ulysses S. Grant:** Gary B. Mills, *Of Men and Rivers: The Story of the Vicksburg District*, U.S. Army Engineer District, Vicksburg, 1978. 28–33.

83 **Twain, who briefly fought for the Confederacy:** "Mark Twain: Staunch Confederate? Once Upon a Time, 150 Years Ago, Baylor Professor Says," blog post, Baylor University, Media and Public Relations, June 7, 2011, https://www.baylor.edu/mediacommunications/news.php?action=story &story=95197#:~:text='%22,the%20Confederate%20cause%20as%20 halfhearted.

84 **Goodin even coauthored a paper:** Robert J. Rapp, Lester Goodin, and Robert Davinroy, "Vegetative Based Solution to Controlling Overbank Scour in the Mississippi River Floodplain Mississippi County, Southeast, Missouri," *Journal of the American Society of Civil Engineers*, 1988.

85 **Though he loved his father, Goodin always felt closer to his uncle:**

A. J. Drinkwater's obituary read: "Survivors include: one *son*, Lester Whitehead Goodin of Cape Girardeau, *son* of Virginia Whitehead Goodin Ridings
of Charleston and the late A. Vernon Goodin" (my italics), https://stan
dard-democrat.com/story/1331517.html.

86 **One hundred million years ago:** Roy B. Van Arsdale and Randel T. Cox,
"The Mississippi's Curious Origins," *Scientific American*, 2007. To craft the
following six paragraphs, I corresponded at some length with Van Arsdale
and with Pat Mulvaney of the Missouri Geological Survey.

87 **which the Anishinaabe people called:** "The Great Lakes: An Ojibwe
Perspective," The Decolonial Atlas, April 14, 2015, https://decolonialatlas
.wordpress.com/2015/04/14/the-great-lakes-in-ojibwe-v2/.

88 **During the winter of 1811–1812:** "Earthquake Hazards, the New Madrid
Seismic Zone, Overview, Science of the New Madrid Seismic Zone" (USGS
.gov), https://www.usgs.gov/natural-hazards/earthquake-hazards/science
/new-madrid-seismic-zone?qt-science_center_objects=0#qt-science_cen
ter_objects.

88 **Eliza Bryan was then thirty-one years old:** For information about Bryan,
and other eyewitness accounts of the 1811–1812 New Madrid earthquakes,
I used the University of Memphis, Center for Earthquake Research and
Information, New Madrid Compendium Eyewitness Accounts, https://
www.memphis.edu/ceri/compendium/eyewitness.php.

88 **Tremors were felt in Boston; people were awakened in Washington,
DC:** Myron L. Fuller, *The New Madrid Earthquake* (U.S. Geological Survey,
Bulletin 494. 1912), 17, 28–30.

88 **Timothy Flint, who passed by a few years later:** Timothy Flint, *Recollections of the Last Ten Years, Passed in Occasional Residences and Journeyings in the
Valley of the Mississippi, from Pittsburg and the Missouri to the Gulf of Mexico,
and from Florida to the Spanish Frontier; in a Series of Letters to the Rev. James
Flint, of Salem, Massachusetts* (Boston: Cummings, Hilliard, and Company,
1826), 222.

89 **The Mississippi sloshed side-to-side in tidal waves:** Fuller, *The New
Madrid Earthquake*, 91–92.

89 **The last great quake, on February 7:** William Clark, of Lewis and Clark
fame, was then governor of Missouri Territory and asked Congress for relief.
He received $50,000, the first federal disaster relief bill.

89 **A 2009 report commissioned by the Federal Emergency Management
Agency:** *Impact of New Madrid Seismic Zone Earthquakes on the Central USA*,
Vol. 1 and 2, MAE Center Report No. 09-03, October 2009.

89 **"the highest economic losses due to a natural disaster in the United
States":** Carey Gillam, "Government Warns of 'Catastrophic' U.S.

Quake," Reuters, November 20, 2008, https://www.reuters.com/article/us-earthquake-study/government-warns-of-catastrophic-u-s-quake-idUSTRE4AJ9EV20081120.

89 **"widespread and catastrophic physical damage":** Ibid. Both quotes come from a FEMA press release characterizing the report's findings.

90 **The destruction and death of 1927 finally convinced the nation to take:** Information for this paragraph and the next comes from Barry, *Rising Tide*; Charles A. Camillo, *Divine Providence: The 2011 Flood in the Mississippi River and Tributaries Project* (Vicksburg: Mississippi River Commission, 2012), 17; and personal communications with Brian Rentfro.

92 **There had been eight breaches near Morganza in the 1800s:** James F. Barnett Jr., *Beyond Control: The Mississippi River's New Channel to the Gulf of Mexico* (Jackson: University Press of Mississippi, 2017), 111.

94 **Even before the waters receded, the *Kansas City Star* sent:** "The Great Flood in Missouri as Seen and Recorded by Thomas Hart Benton," *Kansas City Star*, February 14, 1937. See also: Justin Wolff, *Thomas Hart Benton: A Life* (New York: Farrar, Straus and Giroux, 2012), 283.

94 **Benton wrote about the experience in his autobiography:** Thomas Hart Benton, *An Artist in America* (New York: Robert M. McBride & Company, 1937), 146–147.

95 **The sisters and their extended family lived in the Black community of Pinhook:** I visited the ruins of Pinhook and conducted interviews with Twan, Debra, and Aretha, but in this chapter I also rely on an exhaustively researched book by two sociologists: David Todd Lawrence and Elaine J. Lawless, *When They Blew the Levee: Race, Politics, and Community in Pinhook, Missouri* (Jackson: University Press of Mississippi, 2018).

95 **About fifty years later, their father, Jim Robinson Jr.:** Jim Robinson Jr., interviewed by Will Sarvis, 1998, Pinhook, Missouri, audio recording and transcript, The Oral History Program of the State Historical Society of Missouri, Politics in Missouri Oral History Project, Columbia, Missouri. (Jim Jr. was sixty-five in 1998. He died in 2004 at age seventy.)

95 **Jim Sr. and his partners could buy only:** "When we moved from Tennessee, my people were not allowed to own certain lands and live in town. We were only able to purchase the land that the Mississippi River flooded." Jim Robinson Jr., 2002, Senate testimony.

96 **Once in a while, it made it up the steps:** Ibid.

98 **Then her brother called from New York:** Lawrence and Lawless, *When They Blew the Levee*, 89.

99 **Blowing the levees in 1937 successfully mitigated:** For the history of the MR&T Project and the day-by-day story of the 2011 flood from the

perspective of the Corps, I rely heavily on: Camillo, *Divine Providence*, and personal communication with Camillo.

99   **As the Ohio Valley snowpack melted in the late winter of 2011:** Ibid., 23–24.

101   **Boils attacked the levees downriver in Fulton County:** Ibid., 70.

101   **On April 27, Walsh and his staff attended a town hall meeting:** M. D. Kittle, "Residents Voice Concerns About Breaching Birds Point Levee," *Southeast Missourian*, April 28, 2011.

101   **Guarded and unflappable, Walsh stood there:** Camillo, *Divine Providence*, 74–76.

102   **"That equates to about 2.5 million cubic feet per second":** Kittle, "Residents Voice Concerns About Breaching Birds Point Levee." The rest of the meeting dialogue comes from Kittle's story and personal communications with Kevin Mainord.

104   **Paul Haney and his wife were at home when a deputy knocked on their door:** Paul Haney passed away on June 19, 2019, before I could fact-check these portions of the manuscript. His brother, Ronnie Haney, confirmed the material to the best of his knowledge.

105   **As he went to bed that night:** This paragraph is based on Michael Walsh's recollections as told to me in 2020.

111   **In 2011, Cairo was 70 percent Black:** Cairo was 70 percent Black in 2010, and its poverty rate was 50 percent in 2012, according to data.census .gov.

111   **But the steamboats disappeared, then the railroads:** Tim Murphy, "Donald Trump Asked, 'What Do You Have to Lose?' This Illinois Town Found Out: How a small town got caught up in Ben Carson's crusade against fair housing." *Mother Jones*, July/August 2018.

111   **nearly three thousand residents:** By 2019, Cairo's population had fallen to 2,120, according to the U.S. Census (census.gov).

112   **Scientists from the University of California, Irvine:** Adam Luke et al., "Hydraulic modeling of the 2011 New Madrid Floodway activation: a case study on floodway activation controls." *Natural Hazards* 77, 1863–1887 (2015). https://doi.org/10.1007/s11069-015-1680-3.

113   **Several residents had flood insurance:** Mary Delach Leonard, "Pinhook, Mo.: Levee breach destroyed the village, but changing times had already taken a toll," St. Louis Public Radio, March 11, 2013.

113   **Anthony Schick of the *Columbia Missourian* reported:** Anthony Schick, "Mississippi River Town of Pinhook Struggles to Reclaim Its Community After Levee Break," *Columbia Missourian*, May 5, 2012.

114   **The church and houses stood abandoned, curtains flapping in the**

wind: David Todd Lawrence and Elaine Lawless made a film, too, *Taking Pinhook*, which shows this scene, https://rebuildpinhook.org/documentary/.

114 **In *Damages Done: The Longitudinal Impacts*:** Junia Howell and James R. Elliott, *Damages Done: The Longitudinal Impacts of Natural Hazards on Wealth Inequality in the United States* (Oxford University Press, Society for the Study of Social Problems, 2018).

115 **Again and again in interviews, Debra:** This quote comes from Lawrence and Lawless' website: https://rebuildpinhook.org/ See also: Leonard, "Pinhook, Mo.: Levee breach destroyed the village."

115 **From 1980 to mid-2020, the United States experienced:** NOAA National Centers for Environmental Information (NCEI), U.S. Billion-Dollar Weather and Climate Disasters (2020), https://www.ncdc .noaa.gov/billions/.

115 **Between 2004 and 2015, FEMA distributed:** Rocio Cara Labrador and Amelia Cheatham, "U.S. Disaster Relief at Home and Abroad," Council on Foreign Relations, August 24, 2020, https://www.cfr.org/backgrounder/us -disaster-relief-home-and-abroad.

116 **FEMA had offered to buy out the village for $1.7 million:** "2 Missouri communities weigh possible flood buyouts," The Associated Press, September 2, 2011.

116 **an additional $1.4 million was being sought:** Schick, "Mississippi River Town of Pinhook Struggles to Reclaim Its Community After Levee Break."

118 **FEMA bought out twelve homeowners, plus the church:** The individual Pinhook buyouts are listed in this NPR database under zip code 63845, https://www.npr.org/2019/03/05/696995788/search -the-thousands-of-disaster-buyouts-fema-didnt-want-you-to-see.

118 **Four hundred and fifty-two thousand dollars had come through from HUD:** HUD grant B-12-DT-29-0001, Project # 2012-DT-15, Grantee: Village of Pinhook, https://drgr.hud.gov/public/downloads/action-plans/B -12-DT-29-0001-AP.pdf.

119 **the authorities should have given the residents greater warning and assistance:** Since the 2011 activation, the Corps has set up a dedicated office in East Prairie to be staffed whenever there is a possibility that the floodway might be used.

120 **When the Sacramento River rises to a certain height:** Brad Walker, "Comparison of the Birds Point–New Madrid Floodway, Mississippi River and the Yolo Bypass, Sacramento River," *Journal of Earth Science* 27, no. 1 (2016): 47–54, doi:10.1007/s12583-016-0628-1, http://en.earth-science .net.

126 **Convincing Congress would be even harder:** S.3021, "America's Water

Infrastructure Act of 2018," Public Law 115-270, 115th Congress, https://www.congress.gov/bill/115th-congress/senate-bill/3021/text.

126 **Yet, over time, perception of the standard shifted:** "Investigation of the Performance of the New Orleans Flood Protection Systems in Hurricane Katrina on August 29, 2005," Independent Levee Investigation Team, Final Report, July 31, 2006, section 12: 12.

127 **The Corps considered the flooding of the city:** *A Failure of Initiative: Final Report of the Select Bipartisan Committee to Investigate the Preparation for and Response to Hurricane Katrina* (Washington, DC: U.S. Government Printing Office, 2006).

127 **In 1999 Congress authorized the Corps:** "The Creeping Storm," *Civil Engineering Magazine*, ASCE, June 2003.

127 **"That's a lot of time," said an engineer:** The engineer was Al Naomi. Before, during, and after Katrina, Naomi was the Corps' project manager for hurricane protection levees in Southeast Louisiana. This quote, along with the preceding paragraph, comes from "The Creeping Storm."

127 **In July of 2004 FEMA performed a separate exercise:** "Hurricane Pam Exercise Concludes," FEMA press release, July 23, 2004, Release Number: R6-04-093.

127 **Katrina smashed into Southeast Louisiana's outdated defenses:** "Investigation of the Performance of the New Orleans Flood Protection Systems," xix, xx.

128 **In the Corps' final report on Katrina:** *Performance Evaluation of the New Orleans and Southeast Louisiana Hurricane Protection System*, Final Report of the Interagency Performance Evaluation Task Force, Volume I: Executive Summary and Overview, June 2009, I-6, https://biotech.law.lsu.edu/katrina/ipet/Volume%20I%20FINAL%2023Jun09%20mh.pdf.

128 **The concrete was barely dry before the Corps announced that:** *Federal Register* 84, no. 63 (Tuesday, April 2, 2019): Notices, 12598, https://www.eenews.net/assets/2019/04/11/document_ew_01.pdf. According to the Corps' New Orleans District, the levee lifts have been accomplished, and as of 2020, all the HSDRRS levees were at or above one-hundred-year protection.

128 **Katrina was considered a four-hundred-year storm:** Dan Swenson, "Hurricane Katrina flooding compared to a 500-year storm today: Graphic," *Times-Picayune*, August 16, 2013, https://www.nola.com/news/weather/article_a07212b9-6057-5ed6-8914-07b8135a430b.html.

129 **In 2017 CPRA released a fifty-year, $50 billion Master Plan:** "Louisiana's Comprehensive Master Plan for a Sustainable Coast," Coastal Protection and Restoration Authority of Louisiana, effective June 2, 2017.

PART III—Rivers of Earth

135 **Inquiring after an unfamiliar tributary:** Donald M. Lance, "The Origin and Meaning of 'Missouri,'" *Names: A Journal of Onomastics*: Vol. 47 (1999). The Illinoisan people whom the French explorers interviewed (near present-day Keokuk, Iowa) called the river *Pekitanoui*, literally "it mud-flows." The Illinoisans called the people who lived beside this river something like "Missouri," meaning "people with log canoes."

135 **In his journal of June 21, 1804:** Journal of William Clark, June 21, 1804, https://lewisandclarkjournals.unl.edu/item/lc.jrn.1804-06-21.

135 **Half the Missouri's watershed is semi-arid:** Personal communications with Robert Kelley Schneiders, author of *Unruly River: Two Centuries of Change Along the Missouri* (Lawrence: University Press of Kansas, 1999).

135 **In pre-dam times, the Missouri supplied:** J. S. Alexander, R. C. Wilson, and W. R. Green, *A Brief History and Summary of the Effects of River Engineering and Dams on the Mississippi River System and Delta*, U.S. Geological Survey Circular 1375, 2012.

135 **Spanning the river near Yankton, South Dakota:** Gavins Point Project Statistics, published August 6, 2012, https://www.nwo.usace.army.mil/Media/Fact-Sheets/Fact-Sheet-Article-View/Article/487639/gavins-point-project-statistics/.

138 **According to Boyd, the best work on the subject:** Matthew W. George, S.M.ASCE; Rollin H. Hotchkiss, Ph.D., P.E., D.WRE, F.ASCE; and Ray Huffaker, Ph.D., "Reservoir Sustainability and Sediment Management," *Journal of Water Resources Planning and Management*, ASCE, 2016.

139 **Here, too, the Missouri's bed has:** "Missouri River Bed Degradation Feasibility Study Technical Report," U.S. Army Corps of Engineers, Kansas City District, May 2017. The commercial mining of sand and gravel from the riverbed is also contributing to bed degradation on the Lower Missouri River. The industry has been curtailed in recent years, but not banned. In-river mining permits, issued by the Corps, are up for renewal in 2020.

141 **according to the Corps' 1949 study of the project:** "Preliminary Studies Relating to the Gavins Point Reservoir Project," U.S. Army Corps of Engineers, Omaha District, March 30, 1949.

142 **The brick storefronts and frame houses had vanished:** Mary Pat Murphy, "Niobrara Being Menaced by Dam Built to Save It," (AP) *Argus-Leader* (Sioux Falls, SD), June 6, 1971.

142 **In a history subtitled:** John E. Carter, "Niobrara, Nebraska: The Town Too Tough to Stay Put!," *Nebraska History* 72 (1991): 144–149.

143 **The fine sandy soil, dug up for foundations and yards:** Alan Kemp

told me this. Kemp is a lifelong Niobrara resident and co-owner of Moody Motor, the local Ford dealership. He was also kind enough to fly me over the Lewis and Clark delta in his Cessna Skyhawk II.

143 **If the Niobrara River ever veered from its current bed:** In the flood of 2019 the Niobrara River did cut across the old town site. Ice carried by the flooding Niobrara also pushed the railroad bridge into the river.

143 **The town had lost more than half its population:** According to the U.S. Census (census.gov).

143 **The old state park lay just upstream:** George Vecsey, "Nebraska Town, Slowly Drowning Near Federal Dam, May Move to Higher Ground," *New York Times*, June 17, 1971.

144 **The Santee Tribe was living in Southern Minnesota when:** Minnesota Historical Society, U.S.-Dakota War of 1862 website and resources, https://www.usdakotawar.org/history/war/causes-war.

144 **Returning from a hunt, four Santee men:** Ibid. https://www.usdakotawar.org/history/acton-incident, also personal communication with Bill Convery, Director of Research at the Minnesota Historical Society.

144 **The remaining sixteen hundred Santee:** Ibid., http://www.usdakotawar.org/history/aftermath.

147 **They couldn't touch the principal:** According to Crosley, this is by congressional mandate and is the way the government settled all tribal dam claims. After ten years, the Santee Sioux Nation could collect interest payments, but they could never touch the principal. Crosley didn't know why.

149 **By one estimate, the annual benefit:** National Research Council, *The Missouri River Ecosystem: Exploring the Prospects for Recovery* (Washington, DC: The National Academies Press, 2002), https://doi.org/10.17226/10277. "Commercial navigation traffic had total benefits of $7.0 million in 1995. . . . There are net benefits of less than $3 million annually from commercial traffic."

149 **If the Corps gave up the navigation channel, it could stop:** Ibid. "Relaxing the responsibility to maintain navigation flows would make it demonstrably easier to introduce flows for improving river ecology in that segment."

149 **The pressure to maintain the navigation channel, some in the basin surmised:** Robert Kelley Schneiders, the author of *Unruly River*, said this.

150 **The modern Mississippi flows past Baton Rouge and New Orleans:** Personal communications with James F. Barnett Jr., author of *Beyond Control: The Mississippi River's New Channel to the Gulf of Mexico* (Jackson: University Press of Mississippi, 2017).

151 The Atchafalaya route is much steeper and 173 miles shorter: "Old River Control Brochure," USACE, New Orleans District, 2009, https://www.mvn.usace.army.mil/Portals/56/docs/PAO/Brochures/OldRiverControlBrochure.pdf.

152 In 1987, the *New Yorker* magazine: John McPhee, "Atchafalaya," *New Yorker*, February 23, 1987.

153 "Atchafalaya" was republished: John McPhee, *The Control of Nature* (New York: Farrar, Straus and Giroux, 1989).

154 The impact of more silt and sand going down the Atchafalaya was hard to fathom: Personal communication with Dean Wilson, the Atchafalaya basin keeper.

156 Clancy assured Kaiser that an ambitious assessment: The assessment, currently called OMAR for the Old, Mississippi, Atchafalaya, and Red Rivers, had been funded and was proceeding as of July 2020.

157 During that flood, the Mississippi crevassed a levee at Cabin Teele: John M. Barry, *Rising Tide: The Great Mississippi Flood of 1927 and How It Changed America* (New York: Touchstone/ Simon & Schuster, 1997), 281.

157 Cypress Creek, a small distributary that was blocked off: Ibid., 160.

158 but this one was allowed to go to trial: Charles A. Camillo, *Divine Providence: The 2011 Flood in the Mississippi River and Tributaries Project* (Vicksburg: Mississippi River Commission, 2012), 138.

158 According to Corps lore, the new chief of engineers: Ibid., 148.

158 Jadwin had opposed cutoffs: Brien R. Winkley, "Man-Made Cutoffs on the Lower Mississippi River, Conception, Construction, and River Response," U.S. Army Engineer District, Vicksburg, March 1977.

159 "If you have a bathtub full of water and want to empty it": Douglas L. Weart and Charles G. Holle, "Be Thou at Peace," *Assembly* 29, Association of Graduates, United States Military Academy (1970). (This was an obituary for Ferguson, who died in 1968.)

159 Money already appropriated for the floodway: Martin Reuss, *Designing the Bayous: The Control of Water in the Atchafalaya Basin 1800–1995*, (Alexandria, VA: U.S. Army Corps of Engineers Office of History, 1998), 196–197.

159 He had helped raise the *Maine* from Havana Harbor: "The Man of the 'Maine,'" *Hampton's Magazine* 27, Columbian-Sterling Publishing Company (July 1911).

159 As recently as 2016, a book about the cutoff program: Damon Manders, *The Cutoff Plan: How a Bold Engineering Plan Broke with U.S. Army Corps of Engineers Policy and Saved the Mississippi Valley*, Natural Disaster Research Series, Prediction and Mitigation (Harvest, AL: BISAC), TEC010000, 2016.

159 **In a 1977 paper, Brien R. Winkley:** Winkley, "Man-Made Cutoffs on the Lower Mississippi River."

162 **Wamsley's lab discovered a "hardpoint":** Roger Gaines et al., "Hickman Hardpoint Potamology Study Mississippi River at River Mile 921," USACE Mississippi Valley Division, Mississippi River Geomorphology & Potamology, 2017.

162 **In their reports, the potamologists:** Ibid.; "The Influence of Geology on the Morphologic Response of the Lower Mississippi River," MRG&P Report no. 17, March 2018 (USACE, MVD, ERDC).

163 **The National Oceanic and Atmospheric Administration, responsible for:** Meredith Westington, "Geographic Names Disappear from Charts, but Not from History," NOAA Office of Coast Survey, March 21, 2014. Also, compare NOAA Chart #11358-2018 with 11358-1990.

165 **These days, the Mississippi-Atchafalaya system moves:** Alexander et al., *A Brief History and Summary of the Effects.* Also personal communication with Torbjörn Törnqvist.

166 **The rate of land loss accelerated from forty-three hundred:** Ibid.

171 **"Get John Lopez and go walking":** Al Sunseri, co-owner of P&J Oyster Company, a venerable oyster distributor in the French Quarter, said this to me.

172 **Lopez wasn't bitter; he'd been an architect of the policy:** J. A. Lopez, "The Multiple Lines of Defense to Sustain Coastal Louisiana," *Journal of Coastal Research* Special Publication (2009): 54.

172 **"We will stay as long as it takes":** "Bush: 'We will do what it takes,'" CNN.com, Thursday, September 15, 2005.

173 **A planned retreat would also have required complex expropriation:** Richard Campanella, an architecture professor at Tulane University, talked with me about this moment. Campanella wrote some insightful and balanced columns about the retreat question shortly after the storm. See: Richard Campanella, "In Post-Katrina New Orleans: Abandon? Maintain? Concede?" *Times-Picayune*, April 20, 2006.

173 **Louisiana harvests 850 million pounds:** According to Louisianaseafood.com, the website of the state's Seafood Promotion & Marketing Board.

174 **the promised compensation never came:** Barry, *Rising Tide*, 357. Caernarvon-related claims from St. Bernard and Plaquemines Parish totaled $35 million. The city paid out $2.9 million, mostly to large claimants (Cormier called them "oligarchs"). Of the smaller claimants, 2,809 received $284 each, on average. An additional 1,024 got nothing at all.

175 **In premodern times, when the Mississippi River flooded:** I learned a lot about the history of Plaquemines Parish from Rod Lincoln, a local historian.

177 **In 2017, Mardi Gras Pass was swallowing 1 percent:** By end of the 2019 flood it was 2.5 percent, according to John Lopez.

179 **To figure out how (and if) the diversion would work:** John Schwartz, "A Mini-Mississippi River May Help Save Louisiana's Vanishing Coast," *New York Times*, February 25, 2020.

179 **"white acrylic with the consistency of sugar":** Ibid.

180 **Wamsley hoped CPRA would do this kind of study:** CPRA eventually did do a flume study, though it hasn't released the results.

180 **The oyster farmers and their allies insist:** A lot of these oyster industry generalizations and opinions come from Al Sunseri.

183 **Scientists at Tulane University concluded in a 2020 paper:** Torbjörn E. Törnqvist et al., "Tipping points of Mississippi Delta marshes due to accelerated sea-level rise," *Science Advances*, May 22, 2020: Vol. 6, no. 21.

183 **an interview with Mark Schleifstein of the *Times-Picayune*:** Mark Schleifstein, " 'We're screwed': The Only Question Is How Quickly Louisiana Wetlands Will Vanish, Study Says," *Times-Picayune/New Orleans Advocate*, May 22, 2020.

CODA—Retreat and Fortify

185 **the Netherlands was the place Katrina changed:** Tom Hundley, "Katrina Bolsters Dutch Devotion to New Strategy on Flood Threat," *Chicago Tribune*, April 10, 2006.

185 **In response, their government created the first delta commission:** Deltawerken online, http://www.deltawerken.com/Deltaworks/23.html.

186 **A disaster was still possible:** Robert Slomp, "Flood Risk and Water Management in the Netherlands: A 2012 update," Rijkswaterstaat Ministry of Infrastructure and the Environment, July 9, 2012. Robert Slomp of Rijkswaterstaat clarified many aspects of Dutch flood-control history and policy for me.

186 **"Katrina shook everyone awake," said a spokesman:** Toby Sterling, "Katrina Damage Has Dutch Rethinking Their Flood Plans," Associated Press, November 18, 2005.

186 **The second delta commission presented its program:** "Working Together with Water: A Living Land Builds for Its Future," Deltacommissie, 2008, http://www.deltacommissie.com/doc/deltareport_full.pdf.

186 **In thirty-four places, people were relocated:** "Tailor Made Collaboration: A Clever Combination of Process and Content," Rijkswaterstaat Room for the River, 2013.

187 "Planning for resilience with its inherent uncertainties": Hans de Bruijn et al., *Procedia Computer Science* 44 (2015): 659–668.

187 Room for the River had two stated objectives: "Tailor Made Collaboration."

187 Before Katrina, the Netherlands had budgeted €1 billion annually: "Dutch Draw Up Drastic Measures to Defend Coast Against Rising Seas," *New York Times*, September 3, 2008.

187 While Room for the River cost €2.3 billion: This was the cost of the first phase of Room for the River, which was designed to protect against a maximum Rhine flood of sixteen thousand cubic meters per second (m3/s), up from fifteen thousand. The Delta Program ultimately recommended protection against an eighteen thousand m3/s flood. Increasing the level of protection by such a factor will require considerably more money and more land forfeiture.

189 the Dutch spent on flood protection the same percentage: U.S. national defense spending averaged 5 to 10 percent of annual GDP during the Cold War. The Delta Works and the Zuiderzee Works cost about 6 to 7 percent of Dutch annual GDP during construction. See: Michael E. O'Hanlon, "Is US Defense Spending Too High, Too Low, or Just Right?" Brookings .edu, October 15, 2019, https://www.brookings.edu/policy2020/voter vital/is-us-defense-spending-too-high-too-low-or-just-right/; Frits Bos and Peter Zwaneveld, "Cost-benefit analysis for flood risk management and water governance in the Netherlands: An overview of one century," CPB Netherlands Bureau for Economic Policy Analysis, August 2017.

189 In 2007, the state began receiving 37.5 percent of what the feds collect: Per the Gulf of Mexico Energy Security Act of 2006 (GOMESA) Pub. Law 109-4000, https://www.boem.gov/oil-gas-energy/energy-economics /gulf-mexico-energy-security-act-gomesa.

191 They fear being disliked, delivering bad news, or taking someone's land: After Superstorm Sandy, New Jersey Governor Chris Christie said, "We will rebuild. We will get better and we'll be even tougher." It was the high point of his governorship. https://www.nbcphiladelphia.com/news /local/Gov-Chris-Christie-Press-Conference—1773100091.html.

195 The polder, a piece of reclaimed delta: Herman Havekes et al., "Water Governance: The Dutch Water Authority Model," Dutch Water Authorities, 2017.

196 Perhaps more important than what the Dutch have built is the organizational: Ibid.

196 It is possible to reverse global warming: David Herring and Rebecca Lindsey, "Can We Slow or Even Reverse Global Warming?" Climate.gov,

October 29, 2020, https://www.climate.gov/news-features/climate-qa
/can-we-slow-or-even-reverse-global-warming.

196 **In actuality, experts predict that by the end of this century:** Rebecca
Lindsey, "Climate Change: Global Sea Level," Climate.gov, August 14,
2020, https://www.climate.gov/news-features/understanding-climate
/climate-change-global-sea-level; and: Michon Scott, "Prepare for
More Downpours: Heavy Rain Has Increased Across Most of the
United States, and Is Likely to Increase Further," Climate.gov, July 10,
2019, https://www.climate.gov/news-features/featured-images/pre
pare-more-downpours-heavy-rain-has-increased-across-most-united-0.

About the Author

Tyler J. Kelley was born and raised a few miles from the Mississippi River in Minneapolis. Kelley has written for the *New York Times*, *Wall Street Journal*, and NewYorker.com, among other publications, and currently teaches at The New School in the Journalism + Design program. His previous projects include the documentary film *Following Seas*, codirected with his wife, Araby Kelley. The couple now live in Brooklyn with their son, who has been messing about in boats since he was six weeks old.